Green Energy and Technology

Miklas Scholz

Wetland Systems

Storm Water Management Control

 Springer

Prof. Dr. Miklas Scholz
University of Salford
Discipline of Civil Engineering
School of Computing, Science and Engineering
Faculty of Science, Engineering and Environment
Salford
Greater Manchester
M5 4WT
United Kingdom
m.scholz@salford.ac.uk

ISSN 1865-3529
ISBN 978-1-84996-458-6   e-ISBN 978-1-84996-459-3
DOI 10.1007/978-1-84996-459-3
Springer London Dordrecht Heidelberg New York

British Library Cataloguing in Publication Data
A catalogue record for this book is available from the British Library

Library of Congress Control Number: 2010937239

Cover design: eStudioCalamar, Girona/Berlin

Printed on acid-free paper

Springer is part of Springer Science+Business Media (www.springer.com)

*I would like to dedicate this book
to my wider family
who supported me during my studies
and career.
Particular thanks go to my partner Åsa,
my children Philippa and Jolena,
my mother Gudrun,
my twin sister Ricarda,
and my partner's father Ivan.*

Preface

This book was written in response to the positive feedback received from the wetland community with respect to the book *Wetlands Systems to Control Urban Runoff* by Miklas Scholz and published by Elsevier (Amsterdam) in 2006. Moreover, this new textbook covers all key types of wetland systems and not just those controlling urban runoff. The subtitle "Storm Water Management Control" emphasizes the shift in focus.

The book covers broad water and environmental engineering issues relevant for the drainage and treatment of storm water and wastewater, providing a descriptive overview of complex 'black box' treatment systems and the general design issues involved. The fundamental science and engineering principles are explained to address the student and professional markets. Standard and novel design recommendations for predominantly constructed wetlands and related sustainable drainage systems are provided to account for the interests of professional engineers and environmental scientists. The latest research findings in diffuse pollution and flood control are discussed to attract academics and senior consultants who should recommend the proposed textbook to final-year and postgraduate students, and graduate engineers, respectively.

This original and timely book deals comprehensively not only with the design, operation, maintenance, and water-quality monitoring of traditional and novel wetland systems, but also with the analysis of asset performance and modeling of treatment processes and performances of existing infrastructure predominantly in developed countries, as well as the sustainability and economic issues involved therein.

The book has five chapters. Chapter 1 provides an introduction to wetland systems. It is followed by a comprehensive chapter comprising a diverse selection of wetland case studies, and then by a brief chapter on carbon storage and fluxes within freshwater wetlands. Chapter 4 summarizes wetland systems used in sustainable drainage and flood control applications. The final chapter covers a novel modeling application.

This comprehensive textbook is essential for undergraduate and postgraduate students, lecturers, and researchers in civil and environmental engineering, environmental science, agriculture, and ecological fields of sustainable water management. It is an essential reference for the design, operation, and management of wetlands by engineers and scientists working for the water industry, local authorities, non-governmental organizations, and governmental bodies. Moreover, consulting engineers should be able to apply practical design recommendations and refer to a large variety of practical international case studies including large-scale field studies.

The basic scientific principles should also be of interest to all concerned with constructed environments including town planners, developers, engineering technicians, agricultural engineers, and public health workers. The book was written for a wide readership, but enough hot research topics have been addressed to guarantee the book a long shelf life.

Solutions to pressing water quality problems associated with constructed treatment wetlands, integrated constructed wetlands, farm constructed wetlands and storm water ponds, and other sustainable biological filtration and treatment technologies linked to public health engineering are explained in the book. The case study topics are diverse: wetlands including natural wetlands and constructed treatment wetlands; sustainable water management including sustainable drainage systems and sustainable flood retention basins; and specific applications such as wetlands treating hydrocarbons. The research projects are multidisciplinary, holistic, experimental, and modeling-orientated.

The book is predominantly based on experiences of the author over the last 12 years. Original material, published in more than 80 high-ranking journal papers and 100 conference papers, has been revisited and analyzed. Experience gained as an editorial board member of more than ten peer-reviewed journals ensures that this textbook will contain sufficient material to fill gaps in knowledge and understanding and allow the author to discuss the latest cutting edge research in areas such as integrated constructed wetlands and sustainable flood retention basins.

Acknowledgements

I would like to thank all current and previous members of my Sustainable Water Management Research Group at The University of Edinburgh for their research input. Particular thanks go to Dr. Atif Mustafa, Dr. Rory Harrington, Dr. Åsa Hedmark, Dr. Birol Kayranli, Mr. William McMinn, Dr. Paul Eke, Mrs. Aila Carty, Dr. Kate Heal, Dr. Fabrice Gouriveau, Dr. Kazemi Yazdi, Mr. Nilesh Vyavahare, Mr. Caolan Harrington, Mrs. Qinli Yang, Mr. Paul Carrol, and Mr. Oliver Hofmann for their contributions to research papers supporting this textbook.

I would like to acknowledge the institutions that provided funding for my research relevant for this book, in particular, the government of Ireland, the Royal Academy of Engineering, Alexander von Humboldt Foundation, Monaghan County Council, the European Union, Atlantis Water Management, Teagasc, and The University of Edinburgh.

I am also grateful for the support received by the publishing team at Springer.

About the Author

Prof. Dr. Miklas Scholz, CWem, CEnv, CSci, CEng, FHEA, FCIWEM, FIEMA, FICE, holds a Chair in Civil Engineering at the University of Salford. He was previously a Senior Lecturer in Civil and Environmental Engineering at The University of Edinburgh, which is among the top 20 research universities in the world. Moreover, he is a Visiting Professor at the Department of Land Use and Improvement at the Faculty of Environmental Sciences at the Czech University of Life Sciences, Prague, Czech Republic. He is also a Visiting Professor at the College of Environmental Sciences and Engineering at Nankai University, Tianjin, China. Nankai University is one of the top universities in China.

Since April 2008, Prof. Scholz has been an elected member of the Institute of Environmental Management and Assessment (IEMA) Council (Individual Membership Education category). He was voted first out of seven candidates.

The total recent funding to support Prof. Scholz's research group activity is approx. £1,100,000. More than £960,000 has been awarded to him as a principal investigator over the past 5 years. In addition to this direct financial contribution, he has benefited from approx. £2,000,000 of in-kind funding, which includes a fully monitored wastewater treatment wetland in Ireland.

Prof. Scholz was awarded a Global Research Award and an Industrial Secondmentship by The Royal Academy of Engineering, and two Alexander von Humboldt Fellowships by the Alexander von Humboldt Foundation. He also received research funding from the Natural Environment Research Council, the European Union, Teagasc, Monaghan County Council, Glasgow Council, Alderburgh/Atlantis Water Management, Heidelberg/Hanson Formpave, Triton Electronics, and the National Natural Science Foundation of China.

Prof. Miklas Scholz gained work package leader experience from participating in the EU INTERREG grant Strategic Alliance for Integrated Water Management

Actions (SAWA), which has an overall budget of approx. £6,500,000. He is currently leading the project "Sustainable Flood Retention Basins to Control Flooding and Diffuse Pollution."

Prof. Scholz presently supervises nine PhD students and two research fellows. In the past, he successfully supervised 11 PhD students and 4 four post-doctoral research assistants.

His reputation is predominantly based on approx. 85 peer-reviewed journal paper publications and more than 80 peer-reviewed presentations at national and international conferences. Moreover, he is an active editorial board member of more than ten journals including *Landscape and Urban Planning* and *Wetlands* (associate editor). He is frequently invited to present papers and lead workshops across Europe.

Prof. Scholz has a substantial track record of paper and workshop invitations at national and international conferences and institutions. This is mirrored by the prizes and awards received directly by him and his research team members.

He publishes regularly in *Bioresource Technology* and *Water Research*, which have impact factors of approx. 4.5 and 4.3, respectively. Other papers are regularly published in the *Science of the Total Environment*, *Biotechnology Progress* (and *Sustainable Energy*), and *Ecological Engineering*, which have impact factors of about 2.6, 2.1, and 1.8, respectively.

Prof. Scholz's internationally leading research is in urban and rural runoff control with particular emphasis on solving water quality and climate change problems associated with sustainable flood retention basins, sustainable drainage system technology and planning, constructed treatment wetlands and ponds, and biological filtration technologies linked to public health. Since 1999, he has led more than 20 substantial research projects in sustainable water management, constructed treatment wetlands, and biofiltration.

His research is cutting-edge, multidisciplinary, holistic, and predominantly experimental. However, he has also undertaken comprehensive modeling projects. His research group covers key subject disciplines in engineering and science including sustainable urban water engineering and management, civil and environmental engineering, biotechnology, chemical process technology engineering, and climate change and environmental science. This has led to the successful knowledge transfer of research findings to industry via industrial collaborators such as Hanson Formpave and Triton Electronics. For example, Prof. Scholz has pioneered the combination of permeable pavement systems with ground source heat pumps in sustainable drainage research.

Output from his research group has been featured around 30 times in the national and international media. He is a referee for approx. 60 journals including the top journals in his research area. His national research reputation in urban water technology and management is outstanding, because he has introduced novel design and management guidelines to the engineering community that have been taken up by industry and governmental departments (*e.g.*, Ireland, Northern Ireland, and Scotland).

Contents

Abbreviations

AFTW	=	Aesthetic flood treatment wetland (*SFRB type 4*)
ANN	=	Artificial neural network
ANOVA	=	Analysis of variance
BMP	=	Best management practice (*US phrase for SUDS*)
BMU	=	Best matching unit
BOD	=	Biochemical oxygen demand (*mg/l*); usually 5 d at 20°C, and addition of N-allythiourea
BTEX	=	Benzene, toluene, ethylbenzene, and xylenes (*group of related hydrocarbons*)
COD	=	Chemical oxygen demand (*mg/l*)
DNA	=	Deoxyribonucleic acid
DO	=	Dissolved oxygen (*mg/l or %*)
EDTA	=	Ethylenediaminetetraacetic acid
EU	=	European Union
FCW	=	Farm constructed wetland (*sub-group of ICW*)
GPS	=	Geographical positioning system
HFRB	=	Hydraulic flood retention basin (*SFRB type 1*)
ICW	=	Integrated constructed wetland
IFRW	=	Integrated flood retention wetland (*SFRB type 5*)
MRP	=	Molybdate reactive phosphorus (*mg/l*)
MSE	=	Mean squared error
NFRW	=	Natural flood retention wetland (*SFRB type 6*)
PCA	=	Principal component analysis
PCR	=	Polymerase chain reaction
R^2	=	Coefficient of determination
rRNA	=	Ribosomal ribonucleic acid
SAWA	=	Strategic alliance for water management actions (*name of an EU consortium*)
SD	=	Standard deviation
SFRB	=	Sustainable flood retention basin

SFRW	=	Sustainable flood retention wetland (*SFRB type 3*)
SOM	=	Self-organizing map
SRP	=	Soluble reactive phosphorus (*mg/l*)
spp	=	Species
SS	=	(*Total*) suspended *solids* (*mg/l*)
SUDS	=	Sustainable (*urban*) drainage system
TC	=	Total coliform
TFRB		Traditional flood retention basin (*SFRB type 2*)
U-matrix	=	Unified distance matrix

Chapter 1
Introduction to Wetland Systems

Abstract This introductory chapter provides a brief overview of the key wetland principles, which are not comprehensively covered in subsequent chapters. Most information provided is accepted knowledge that has been widely published elsewhere. The fundamental hydrological, physical, and biochemical processes within wetland systems are reviewed briefly. The relationships between aggregates and microbial and plant communities, as well as the reduction of predominantly biochemical oxygen demand, suspended solids, and heavy metals, are investigated. Most constructed wetland research studies show that after maturation of the biomass that dominates the litter zone, organic and inorganic contaminants are usually reduced similarly for all constructed wetland types. This finding is, however, still controversial, and further research needs to be undertaken. Particular emphasis in the introduction is given to treatment wetlands and wetlands used as sustainable drainage systems to control urban runoff. These technologies are further discussed with the help of recent and relevant research case studies in subsequent chapters.

1.1 Background

Wetlands have been recognized as a valuable natural resource throughout human history. Their importance has been appreciated in their natural state by such people as the Marsh Arabs around the confluence of the rivers Tigris and Euphrates in southern Iraq, as well as in managed forms; *e.g.*, rice paddies, particularly in southeast Asia (Scholz 2006).

The water purification capability of wetlands is being recognized as an attractive option in wastewater treatment. For example, the UK Environment Agency spends significant amounts of money on reed bed schemes in England and Wales. Such systems are designed, for example, to clean up mine water from collieries on which constructed wetlands and associated community parks are being built.

M. Scholz, *Wetland Systems*
© Springer 2011

Figure 1.1 Irish wetland example with high biodiversity value

Reed beds provide a useful complement to traditional sewage treatment systems. They are often a cheap alternative to expensive wastewater treatment technologies such as trickling filters and activated sludge processes. Vertical-flow and horizontal-flow wetlands based on soil, sand, or gravel are used to treat domestic and industrial wastewater. They are also applied to passive treatment of diffuse pollution including mine drainage as well as urban and motorway runoff after storm events.

Furthermore, wetlands serve as a wildlife conservation resource and can also be seen as natural recreational areas for the local community. *Phragmites* spp., *Typha* spp., and other macrophytes typical for swamps are widely used in Europe and North America. Figures 1.1 and 1.2 show representative wetlands with a high biodiversity value.

Figure 1.2 Spanish wetland with high biodiversity value

1.2 Definitions

Defining wetlands has long been a problematic task, partly due to the diversity of environments, which are permanently or seasonally influenced by water, but also due to the specific requirements of the diverse groups of people involved with the study and management of these habitats. The Ramsar Convention, which brought wetlands to the attention of the international community, proposed the following definition (Convention on Wetlands of International Importance Especially as Waterfowl Habitat 1971): "Wetlands are areas of marsh, fen, peatland or water, whether natural or artificial, permanent or temporary, with water that is static or flowing, fresh, brackish or salt, including areas of marine water, the depth of which at low tide does not exceed six meters."

Another, more succinct, definition is as follows (Smith 1980): "Wetlands are a half-way world between terrestrial and aquatic ecosystems and exhibit some of the characteristics of each."

This complements the Ramsar description since wetlands are at the interface between water and land. This concept is particularly important in areas where wetlands may only be 'wet' for relatively short periods of time during a year, such as in areas of the tropics with marked wet and dry seasons.

These definitions put an emphasis on the ecological importance of wetlands. However, the natural water purification processes occurring within these systems have become increasingly relevant to those people involved with the practical use of constructed or even semi-natural wetlands for water and wastewater treatment. There is no single accepted ecological definition of wetlands. However, wetlands are usually characterized by the presence of water, unique soils that differ from upland soils, and the presence of vegetation adapted to saturated conditions.

Whichever definition is adopted, it can be seen that wetlands encompass a wide range of hydrological and ecological types, from high-altitude river sources to shallow coastal regions, in each case being affected by prevailing climatic conditions. For the purpose of this text, however, the main emphasis will be upon wetland systems in a temperate climate.

1.3 Hydrology of Wetlands

The biotic status of a wetland is intrinsically linked to the hydrological factors by which it is affected. These affect the nutrient availability as well as physico-chemical variables such as soil and water pH, and anaerobiosis within soils. In turn, biotic processes will have an impact upon the hydrological conditions of a wetland.

Water is the hallmark of wetlands. Therefore, it is not surprising that the input and output of water (*i.e.*, water budget; see below) of these systems determine the biochemical processes occurring within them. The net result of the water budget (hydroperiod) may show great seasonal variations, but ultimately delineates wet-

Figure 1.3 Spanish wetland that regularly dries out during summer

lands from terrestrial and fully aquatic ecosystems. Wetlands in warm climates are in danger of drying out (Figure 1.3). This is likely to become an increasing problem if climate change can not be mitigated.

From an ecological point of view, as well as an engineering one, the importance of hydrology cannot be overstated, as it defines the species diversity, productivity and nutrient cycling of specific wetlands. That is to say, hydrological conditions must be considered, if one is interested in the species richness of flora and fauna, or if the interest lies in utilizing wetlands for pollution control.

The stability of particular wetlands is directly related to their hydroperiod; that is the seasonal shift in surface and sub-surface water levels. The terms flood duration and flood frequency refer to wetlands that are not permanently flooded and give some indication of the time period involved in which the effects of inundation and soil saturation will be most pronounced.

Of particular relevance to riparian wetlands is the concept of flooding pulses. These pulses cause the greatest difference in high and low water levels, and benefit wetlands by the inflow of nutrients and washing out of waste matter. These sudden and high volumes of water can be observed on a periodic or seasonal basis. It is particularly important to appreciate this natural fluctuation and its effects, since wetland management often attempts to control the level by which waters rise and fall. Such manipulation might be due to the overemphasis placed on water and its role in the lifecycles of wetland flora and fauna, without considering the fact that such species have evolved in such an unstable environment.

In general terms, wetlands are most widespread in those parts of the world where precipitation exceeds water loss through evapotranspiration and surface runoff. The contribution of precipitation to the hydrology of a wetland is influenced by a number of factors. Precipitation such as rain and snow often passes through a canopy of vegetation before it becomes part of the wetland. The volume of water retained by this canopy is termed interception. Variables such as precipitation intensity and vegetation type will affect interception, for which median values of several studies have been calculated as about 10% for deciduous forests and roughly 30% for coniferous woodland. The precipitation that remains to reach the wetland is termed the through-fall. This is added to the stem-flow, which is the water running down vegetation stems and trunks, generally considered a minor component of a wetland water budget (Scholz 2006).

1.4 Wetland Chemistry

Because wetlands are associated with waterlogged soils, the concentration of oxygen within sediments and the overlying water is of critical importance. The rate of oxygen diffusion into water and sediment is slow, and this (coupled with microbial and animal respiration) leads to near anaerobic sediments within many wetlands. These conditions favor rapid peat build-up since decomposition rates and inorganic content of soils are low. Furthermore, the lack of oxygen in such conditions affects the aerobic respiration of plant roots and influences plant nutrient availability. Wetland plants have consequently evolved to be able to exist in anaerobic soils.

While the deeper sediments are generally anoxic, a thin layer of oxidized soil usually exists at the soil–water interface. The oxidized layer is important, since it permits the oxidized forms of prevailing ions to exist. This is in contrast to the reduced forms occurring at deeper levels of soil. The state of reduction or oxidation of iron, manganese, nitrogen, and phosphorus ions determines their role in nutrient availability and also toxicity. The presence of oxidized ferric iron gives the overlying wetland soil a brown coloration, while reduced sediments have undergone 'gleying,' a process by which ferrous iron gives the sediment a blue-grey tint (Scholz 2006).

Therefore, the level of reduction of wetland soils is important in understanding the chemical processes that are most likely to occur in the sediment and influence the water column above. The most practical way to determine the reduction state is by measuring the redox potential, also called the oxidation-reduction potential, of the saturated soil or water. The redox potential quantitatively determines whether a soil or water sample is associated with a reducing or oxidizing environment. Reduction is the release of oxygen and gain of an electron (or hydrogen), while oxidation is the reverse; i.e., the gain of oxygen and loss of an electron.

Oxidation (and therefore decomposition) of organic matter (a much reduced material) occurs in the presence of any electron acceptor, particularly oxygen, but the rate will be slower in comparison with oxygen. Redox potentials are affected by pH and temperature, which influences the range at which particular reactions occur.

Organic matter within wetlands is usually degraded by aerobic respiration or anaerobic processes (e.g., fermentation and methanogenesis). Anaerobic degradation of organic matter is less efficient than decomposition occurring under aerobic conditions.

Fermentation is the result of organic matter acting as the terminal electron acceptor (instead of oxygen as in aerobic respiration). This process forms low-molecular-weight acids (*e.g.*, lactic acid), alcohols (*e.g.*, ethanol), and carbon dioxide. Therefore, fermentation is often central in providing further biodegradable substrates for other anaerobic organisms in waterlogged sediments.

The sulfur cycle is linked with the oxidation of organic carbon in some wetlands, particularly in sulfur-rich coastal systems. Low-molecular-weight organic compounds that result from fermentation (*e.g.*, ethanol) are utilized as organic substrates by sulfur-reducing bacteria during the conversion of sulfate to sulfide.

The prevalence of anoxic conditions in most wetlands has led to their playing a particularly important role in the release of gaseous nitrogen from the lithosphere and hydrosphere to the atmosphere through denitrification. However, the various oxidation states of nitrogen within wetlands are also important to the biogeochemistry of these environments.

Nitrates are important terminal electron acceptors after oxygen, making them relevant in the process of oxidation of organic matter. The transformation of nitrogen within wetlands is strongly associated with bacterial action. The activity of particular bacterial groups is dependent on whether the corresponding zone within a wetland is aerobic or anaerobic.

Within flooded wetland soils, mineralized nitrogen occurs primarily as ammonium, which is formed through ammonification, the process by which organically bound nitrogen is converted into ammonium under aerobic or anaerobic conditions. Soil-bound ammonium can be absorbed through rhizome and root systems of macrophytes and reconverted into organic matter, a process that can also be performed by anaerobic microorganisms.

The oxidized top layer present in many wetland sediments is crucial in preventing the excessive build-up of ammonium. A concentration gradient will be established between the high concentration of ammonium in the lower reduced sediments and the low concentration in the oxidized top layer. This may cause a passive flow of ammonium from the anaerobic to the aerobic layer, where microbiological processes convert the ion into other forms of nitrogen.

In some wetlands, nitrogen may be derived through nitrogen fixation. In the presence of the enzyme nitrogenase, nitrogen gas is converted into organic nitrogen by organisms such as aerobic or anaerobic bacteria and cyanobacteria (blue-green algae). Wetland nitrogen fixation can occur in the anaerobic or aerobic soil layer, overlying water, rhizosphere of plant roots, and on leaf or stem surfaces. Cyanobacteria may contribute significantly to nitrogen fixation.

In wetland soils, phosphorus occurs as soluble or insoluble, organic or inorganic complexes. Its cycle is sedimentary rather than gaseous (as with nitrogen) and predominantly forms complexes within organic matter in peatlands or inorganic sediments in mineral soil wetlands. Most of the phosphorus load in streams and rivers may be present in particulate inorganic form.

Soluble reactive phosphorus (SRP) is the analytical term given to biologically available ortho-phosphate, which is the primary inorganic form. The availability of phosphorus to plants and microconsumers is limited. Under aerobic conditions, insoluble phosphates are precipitated with ferric iron, calcium, and aluminum. Phosphates are adsorbed onto clay particles, organic peat, and ferric and aluminum hydroxides and oxides. Furthermore, phosphorus is bound-up in organic matter through incorporation into bacteria, algae, and vascular macrophytes (Scholz 2006).

There are three general conclusions about the tendency of phosphorus to precipitate with selected ions. In acid soils, phosphorus is fixed as aluminum and iron phosphates. In alkaline soils, phosphorus is bound by calcium and magnesium. Finally, the bioavailability of phosphorus is greatest at neutral to slightly acid pH.

The phosphorus availability is altered under anaerobic wetland soil conditions. The reducing conditions that are typical of flooded soils do not directly affect phosphorus. However, the association of phosphorus with other elements that undergo reduction has an indirect effect upon the phosphorus in the environment. For example, as ferric iron is reduced to the more soluble ferrous form, phosphorus as ferric phosphate is released into solution. Phosphorus may also be released into solution by a pH change brought about by organic, nitric, or sulfuric acids produced by chemosynthetic bacteria. Phosphorus sorption to clay particles is greatest under strongly acidic to slightly acidic conditions.

The physical, chemical, and biological characteristics of a wetland system affect the solubility and reactivity of different forms of phosphorus. Phosphate solubility is regulated by temperature, pH, redox potential, the interstitial soluble phosphorus level, and microbial activity.

Where agricultural land has been converted into wetlands, there can be a tendency towards solubilization of residual fertilizer phosphorus, which results in a rise in the soluble phosphorus concentration in floodwater. This effect can be reduced by physico-chemical amendment, applying chemicals such as alum and calcium carbonate to stabilize the phosphorus in the sediment of these new wetlands.

The redox potential has a significant impact on the dissolved reactive phosphorus of chemically amended soils. The redox potential can vary with fluctuating water table levels and hydraulic loading rates. Dissolved phosphorus concentrations are relatively high under reduced conditions and decrease with increasing redox potential. Iron compounds are particularly sensitive to the redox potential, resulting in the chemical amendment of wetland soils. Furthermore, alum and calcium carbonate are suitable to bind phosphorus even during fluctuating redox potentials.

Macrophytes assimilate phosphorus predominantly from deep sediments, thereby acting as nutrient pumps. The most important phosphorus retention pathway in wetlands is via physical sedimentation. Most phosphorus taken from sediments by macrophytes is reincorporated into the sediment as dead plant material, and therefore remains in the wetland indefinitely. Macrophytes can be harvested as a means to enhance phosphorus removal in wetlands. When macrophytes are harvested at the end of the growing season, phosphorus can be removed from the internal nutrient cycle within wetlands (Scholz 2006).

Moreover, models showed a phosphorus removal potential of three-quarters of that of the phosphorus inflow. Therefore, harvesting would reduce phosphorus levels in upper sediment layers and drive phosphorus movement into deeper layers, particularly the root zone. In deep layers of sediment, the phosphorus sorption capacity increases along with a lower desorption rate.

In wetlands, sulfur is transformed by microbiological processes and occurs in several oxidation stages. Reduction may occur if the redox potential is slightly negative. Sulfides provide the characteristic 'bad egg' odor of wetland soils.

Assimilatory sulfate reduction is accomplished by obligate anaerobes such as *Desulfovibrio* spp. Bacteria may use sulfates as terminal electron acceptors in anaerobic respiration in a wide pH range, but they are highest around neutral. Furthermore, the greatest loss of sulfur from freshwater wetland systems to the atmosphere is via hydrogen sulfide.

Oxidation of sulfides to elemental sulfur and sulfates can occur in the aerobic layer of some soils and is carried out by chemoautotrophic (*e.g.*, *Thiobacillus* spp.) and photosynthetic microorganisms. *Thiobacillus* spp. may gain energy from the oxidation of hydrogen sulfide to sulfur and, further, by certain other species of the genus, from sulfur to sulfate.

In the presence of light, photosynthetic bacteria, such as purple sulfur bacteria of salt marshes and mud flats, produce organic matter. This is similar to the familiar photosynthesis process, except that hydrogen sulfide is used as the electron donor instead of water.

The direct toxicity of free sulfide in contact with plant roots has been noted. There is a reduced toxicity and availability of sulfur for plant growth if it precipitates with trace metals. For example, the immobilization of zinc and copper by sulfide precipitation is well known.

The input of sulfates to freshwater wetlands, in the form of Aeolian dust or as anthropogenic acid rain, can be significant. Sulfate deposited on wetland soils may undergo dissimilatory sulfate reduction by reaction with organic substrates.

Protons consumed during this reaction generate alkalinity. This leads to an increase in pH with depth in wetland sediments. It has been suggested that this 'alkalinity effect' can act as a buffer in acid-rain-affected lakes and streams.

The sulfur cycle can vary greatly within different zones of a particular wetland. The variability in the sulfur cycle within the watershed can affect the distribution of reduced sulfur stored in soil. This change in local sulfur availability can have marked effects upon stream water over short distances.

1.5 Wetland System Mass Balance

The general mass balance for a wetland system, in terms of chemical pathways, uses the following main pathways: inflows, intra-system cycling, and outflows. The inflows are mainly through hydrologic pathways such as precipitation (particularly urban), surface runoff, and groundwater. The photosynthetic fixation of both atmospheric carbon and nitrogen is an important biological pathway. Intra-system cycling is the movement of chemicals in standing stocks within wetlands, such as litter production and remineralization. Translocation of minerals within plants is an example of the physical movement of chemicals. Outflows involve hydrologic pathways, but also include the loss of chemicals to deeper sediment layers, beyond the influence of internal cycling (although the depth at which this threshold occurs is not certain). Furthermore, the nitrogen cycle plays an important role in outflows, such as nitrogen gas lost as a result of denitrification. However, respiratory loss of carbon is also an important biotic outflow.

There is great variation in the chemical balance from one wetland to another. Wetlands act as sources, sinks, or transformers of chemicals depending on wetland type, hydrological conditions, and length of time the wetland has received chemical inputs. As sinks, the long-term sustainability of this function is associated with hydrologic and geomorphic conditions as well as the spatial and temporal distribution of chemicals within wetlands.

Particularly in temperate climates, seasonal variation in nutrient uptake and release is expected. Chemical retention will be greatest in the growing seasons (spring and summer) due to higher rates of microbial activity and macrophyte productivity.

The ecosystems connected to wetlands affect and are affected by the adjacent wetland. Upstream ecosystems are sources of chemicals, while those downstream may benefit from the export of certain nutrients or the retention of particular chemicals.

Nutrient cycling in wetlands differs from that in terrestrial and aquatic systems. More nutrients are associated with wetland sediments than with most terrestrial soils, while benthic aquatic systems have autotrophic activity, which relies more on nutrients in the water column than in sediments.

The ability of wetland systems to remove anthropogenic waste is not limitless. It follows that treatment wetlands may become saturated with certain contaminants and act rather as sources than sinks of corresponding pollutants.

1.6 Macrophytes in Wetlands

Wetland plants are often central to wastewater treatment wetlands. The following requirements of plants should be considered for use in such systems:

- ecological adaptability;
- tolerance of local conditions in terms of climate, pests, and diseases;
- tolerance of pollutants;
- resilience to hypertrophic waterlogged conditions;
- ready propagation, rapid establishment, spread, and growth;
- high pollutant removal capacity; and
- direct assimilation or indirect enhancement of nitrification, denitrification, and other microbial processes.

Interest in macrophyte systems for sewage treatment by the water industry dates back to the 1980s. The ability of different macrophyte species and their assemblages within systems to most efficiently treat specific wastewaters has been proven. The dominant species of macrophyte varies from locality to locality. The number of genera (*e.g.*, *Phragmites* spp., *Typha* spp., and *Scirpus* spp.) common to all temperate locations is great.

The improvement of water quality with respect to key water quality variables including BOD, COD, total SS, nitrates, and phosphates has been proven. However, relatively little work has been conducted on the enteric bacteria removal capability of macrophyte systems.

There have been many studies conducted to determine the primary productivity of wetland macrophytes, although estimates have generally tended to be fairly high. Little of aquatic plant biomass is consumed as live tissue; it rather enters the pool of particulate organic matter following tissue death. The breakdown of this material is consequently an important process in wetlands and other shallow aquatic habitats. Litter breakdown has been studied along with intensive work on *P. australis*, one of the most widespread aquatic macrophytes.

There has been an emphasis on studying the breakdown of aquatic macrophytes in a way that most closely resembles that of natural plant death and decomposition, principally by not removing plant tissue from macrophyte stands. Many species of freshwater plants exhibit the so-called 'standing-dead' decay, which describes the observation of leaves remaining attached to their stems after senescence and death. Different fractions (leaf blades, leaf sheaths, and culms) of *P. australis* differ greatly in structure and chemical composition and may exhibit different breakdown rates, patterns, and nutrient dynamics.

P. australis (Cav.) Trin. ex Steud. (common reed), formerly known as *Phragmites communis* (Norfolk reed), is a member of the large family Poaceae (roughly 8000 species within 785 genera). The common reed occurs throughout most of Europe and is distributed worldwide. It may be found in permanently flooded soils of still or slowly flowing water. This emergent plant is usually firmly rooted in wet sediment but may form lightly anchored rafts of 'hover reed.'

It tends to be replaced by other species at drier sites. The density of this macrophyte is reduced by grazing (*e.g.*, consumption by waterfowl) and may then be replaced by other emergent species such as *Phalaris arundinacea* L. (reed canary grass).

P. australis is a perennial, and its shoots emerge in spring. Hard frost kills these shoots, illustrating the tendency for reduced vigor towards the northern end of its distribution. The hollow stems of the dead shoots in winter are important in transporting oxygen to the relatively deep rhizosphere.

Reproduction in closed stands of this species is mainly by vegetative spread, although seed germination enables the colonization of open habitats. Detached shoots often survive and regenerate away from the main stand.

The common reed is most frequently found in nutrient-rich sites and absent from the most oligotrophic zones. However, the stems of this species may be weakened by nitrogen-rich water and are consequently more prone to wind and wave damage.

Typha latifolia L. (cattail, reedmace, or bulrush) is a species belonging to the small family Typhaceae. This species is widespread in temperate parts of the northern hemisphere, but extends to South Africa, Madagascar, Central America, and the West Indies, and has been naturalized in Australia. *T. latifolia* is typically found in shallow water or on exposed mud at the edges of lakes, ponds, canals, and ditches and less frequently near fast flowing water. This species rarely grows at water depths of >0.3 m, where it is frequently replaced (outcompeted) by *P. australis*.

Reedmace is a shallow-rooted perennial producing shoots throughout the growing season that subsequently die in autumn. Colonies of this species expand by rhizomatous growth at rates of about 4 m/a, while detached portions of rhizome may float and establish new colonies. In contrast, colony growth by seeds is less likely. Seeds require moisture, light, and relatively high temperatures to germinate, although this may occur in anaerobic conditions. Where light intensity is low, germination is stimulated by temperature fluctuation.

1.7 Physical and Biochemical Parameters

The key physico-chemical parameters relevant for wetland systems include the BOD, turbidity, and the redox potential. The BOD is an empirical test to determine the molecular oxygen used during a specified incubation period (usually 5 d) for the biochemical degradation of organic matter (carbonaceous demand) and the oxygen used to oxidize inorganic matter (*e.g.*, sulfides and ferrous iron). An extended test (up to 25 d) may also measure the amount of oxygen used to oxidize reduced forms of nitrogen (nitrogenous demand), unless this is prevented by an inhibitor chemical.

Turbidity is a measure of the cloudiness of water, caused predominantly by suspended material such as clay, silt, organic and inorganic matter, plankton and other microscopic organisms, and scattering and absorbing light. Turbidity in wetlands and lakes is often due to colloidal or fine suspensions, while in fast-flowing waters, the particles are larger and turbid conditions are prevalent predominantly during floods.

The redox potential is another important water quality control parameter for monitoring wetlands. The reactivities and mobilities of elements such as iron, sulfur, nitrogen, carbon, and a number of metallic elements depend strongly on the redox potential conditions. Reactions involving electrons and protons are pH- and redox potential-dependent. Chemical reactions in aqueous media can often be characterized by pH and the redox potential together with the activity of dissolved chemical species.

The redox potential is a measure of intensity and does not represent the capacity of the system for oxidation or reduction. The interpretation of the redox potential values measured in the field is limited by a number of factors including irreversible reactions, 'electrode poisoning,' and multiple redox couples.

1.8 Constructed Treatment Wetlands

Natural wetlands usually improve the quality of water passing through them, effectively acting as ecosystem filters. In comparison, most constructed wetlands are artificially created wetlands used to treat water pollution in its variety of forms. Therefore, they fall into the category of constructed treatment wetlands. Treatment wetlands are solar-powered ecosystems. Solar radiation varies diurnally, as well as on an annual basis.

Constructed wetlands are designed for the purpose of removing bacteria, enteric viruses, SS, BOD, nitrogen (predominantly as ammonia and nitrate), metals, and phosphorus. Two general types of constructed wetlands are usually commissioned in practice: surface-flow (*i.e.*, horizontal-flow) and sub-surface-flow (*e.g.*, vertical-flow) wetlands.

Surface-flow constructed wetlands most closely mimic natural environments and are usually more suitable for wetland species because of permanent standing water. In sub-surface-flow wetlands, water passes laterally through a porous medium (usually sand and gravel) with a limited number of macrophyte species. These systems often have no standing water.

Constructed treatment wetlands can be built at, above, or below the existing land surface if an external water source is supplied (*e.g.*, wastewater). The grading of a particular wetland in relation to the appropriate elevation is important for the optimal use of the wetland area in terms of water distribution. Soil type and groundwater level must also be considered if long-term water shortage is to be avoided. Liners can prevent excessive desiccation, particularly where soils have a high permeability (*e.g.*, sand and gravel) or where there is limited or periodic flow.

Rooting substrate is also an important consideration for the most vigorous growth of macrophytes. A loamy or sandy topsoil layer between 0.2 and 0.3 m in depth is ideal for most wetland macrophyte species in a surface-flow wetland. A sub-surface-flow wetland will require coarser material such as gravel or coarse sand.

High levels of dissolved organic carbon may enter water supplies where soil aquifer treatment is used for groundwater recharge, as the influent for this method is likely to come from long hydraulic retention time wetlands. Consequently, there is a greater potential for the formation of disinfection byproducts.

A shorter hydraulic retention time will result in less dissolved organic carbon leaching from plant material compared to a longer hydraulic retention time in a wetland. Furthermore, dissolved organic carbon leaching is likely to be most significant in wetlands designed for the removal of ammonia, which requires a long hydraulic retention time.

A more detailed discussion on constructed treatment wetlands is beyond the scope of this introduction section. Readers may wish to consult other textbooks such as Scholz (2006).

1.9 Constructed Wetlands Used for Storm Water Treatment

Urban runoff comprises the storm water runoff from urban areas such as roads and roofs. Road and car park runoff is often more contaminated with heavy metals and organic matter than roof runoff. Runoff is usually collected in gully pots that require regular cleaning. Storm water pipes transfer urban runoff either directly to watercourses or to sustainable (urban) drainage systems (SUDS, also called best management practice (BMP) in the USA) such as ponds or wetlands. However, urban runoff is frequently only cleaned by silt traps that need to be cleaned regularly.

Conventional storm water systems are designed to dispose of rainfall runoff as quickly as possible. This results in 'end of pipe' solutions that often involve the provision of large interceptor and relief sewers, huge storage tanks in downstream locations, and centralized wastewater treatment (Scholz 2006).

Storm runoff from urban areas has been recognized as a major contributor to the pollution of the corresponding receiving urban watercourses. The principal pollutants in urban runoff are BOD, SS, heavy metals, deicing salts, hydrocarbons, and fecal coliforms.

In contrast, SUDS such as combined attenuation pond and infiltration basin systems can be applied as cost-effective local 'source control' drainage solutions, *e.g.*, for collection of road runoff. It is often possible to divert all road runoff for infiltration or storage and subsequent recycling. As runoff from roads is a major contributor to the quantity of surface water requiring disposal, this is a particularly beneficial approach where suitable ground conditions prevail. Furthermore, infiltration of road runoff can reduce the concentration of diffuse pollutants such as oil, leaves, metals, and feces, thereby improving the quality of surface water runoff (Scholz 2006).

Although various conventional methods have been applied to treat storm water, most technologies are not cost-effective or are too complex. Constructed wetlands integrated into a BMP concept are a sustainable means of treating storm water and have proven to be more economical (*e.g.*, construction and maintenance) and energy efficient than traditional centralized treatment systems. Furthermore, wetlands enhance biodiversity and are less susceptible to variations of loading rates.

Contrary to standard domestic wastewater treatment technologies, storm water (*e.g.*, gully pot liquor and effluent) treatment systems have to be robust to highly variable flow rates and water quality variations. The storm water quality depends on the load of pollutants present on the road and the corresponding dilution by each storm event.

In contrast to standard horizontal-flow constructed treatment wetlands, vertical-flow wetlands are flat and intermittently flooded and drained, allowing air to refill the soil pores within the bed. While it has been recognized that vertical-flow constructed wetlands usually have higher removal efficiencies with respect to organic pollutants and nutrients in comparison to horizontal-flow wetlands, denitrification is less efficient in vertical-flow systems. When the wetland is dry, oxygen (as part

Figure 1.4 Experimental ponds located in Edinburgh for treatment of road runoff

of the air) can enter the top layer of debris and sand. The following incoming flow of runoff will absorb the gas and transport it to the anaerobic bottom of the wetland (Scholz 2006).

Heavy metals within storm water are associated with fuel additives, car body corrosion, and tire and brake wear. Common metal pollutants from cars include copper, nickel, lead, zinc, chromium, and cadmium. Metals occur in soluble, colloidal, or particulate forms. Heavy metals are most bioavailable when they are soluble, either in ionic or weakly complexed form.

There have been many studies on the specific filter media within constructed wetlands to treat heavy metals economically, *e.g.*, limestone, lignite, activated carbon, peat, and leaves. Metal bioavailability and reduction are controlled by chemical processes including acid volatile sulfide formation and organic carbon binding and sorption in reduced sediments of constructed wetlands. It follows that metals usually accumulate in the top layer (fine aggregates, sediment, and litter) of vertical-flow and near the inlet of horizontal-flow constructed treatment wetlands.

Physical and chemical properties of the wetland soil and aggregates affecting metal mobilization include particle size distribution (texture), redox potential, pH, organic matter, salinity, and the presence of inorganic matter such as sulfides and carbonates. The cation exchange capacity of maturing wetland soils and sediments tends to increase as the texture becomes finer because more negatively charged binding sites are available. Organic matter has a relatively high proportion of negatively charged binding sites. Salinity and pH can influence the effectiveness of the cation exchange capacity of soils and sediments because the negatively charged binding sites will be occupied by a high number of sodium or hydrogen cations.

Sulfides and carbonates may combine with metals to form relatively insoluble compounds. In particular, the formation of metal sulfide compounds may provide

Figure 1.5 Experimental ponds located near Bradford, UK, for treating roof runoff

long-term heavy metal removal, because these sulfides will remain permanently in the wetland sediments as long as they are not reoxidized.

Combined wetlands and infiltration basins are cost-effective 'end of pipe' drainage solutions that can be applied to local source control. The aim is often to assess constraints associated with the design and operation of these systems, the influence of wetland plants on infiltration rates, and the water treatment potential. Road runoff may first be stored and treated in a constructed wetland before it is overflowed into parallel infiltration basins of which one can be planted and the other could be left unplanted (Figure 1.4).

Combined wet and dry ponds as cost-effective 'end of pipe' drainage solutions can be applied to local source control; *e.g.*, diversion or collection of roof drainage (Figure 1.5). It is often possible to divert all roof drainage for infiltration or storage and subsequent recycling. As runoff from roofs is a major contributor to the quantity of surface water requiring disposal, this is a particularly beneficial approach where suitable ground conditions prevail.

Furthermore, roof runoff water is usually considered to be cleaner than road runoff water. However, diffuse pollution from the atmosphere and degrading building materials can have a significant impact on the water quality of any surface water runoff.

Gully pots can be viewed as simple physical, chemical, and biological reactors. They are particularly effective in retaining suspended solids. Currently, gully pot effluent is extracted once or twice per annum from road drains and transported (often over long distances) for treatment at sewage works. A more sustainable solution would be to treat gully pot effluent locally in potentially sustainable constructed wetlands, reducing transport and treatment costs. Furthermore, gully pot effluent treated with constructed wetlands can be recycled for the cleansing of gully pots and washing of wastewater tankers.

Heavy metals within urban runoff are associated with fuel additives, car body corrosion, and tire and brake wear. Common metal pollutants from road vehicles include lead, zinc, copper, chromium, nickel, and cadmium.

Storm water runoff is usually transferred within a combined sewer (rainwater and sewage) or separate storm water pipeline. Depending on the degree of pollution, it may require further conventional treatment at the wastewater treatment plant. Alternatively, storm water runoff may be treated by more sustainable technology including storm water ponds or simply by disposing of it via ground infiltration or drainage into a local watercourse (Scholz 2006).

Common preliminary treatment steps for storm water that is to be disposed into a local watercourse are extended storage within the storm water pipeline and transfer through a silt trap (also referred to as a sediment trap). The silt trap is often the only active preliminary treatment of surface water runoff despite the fact that storm water is frequently contaminated with heavy metals and hydrocarbons. Heavy metals within road runoff are associated with fuel additives, car body corrosion, and tire and brake wear (Scholz 2006).

The dimensioning of silt traps is predominantly based upon the settlement properties of the particles and the maximum flow-through velocity. As a general rule, the outflow should be located on the same axis as the inflow, so that continuous flow will ensure particle settlement over the full length of the silt trap (Schmitt et al. 1999). The structure should be small and simple to keep capital costs low.

Various international and British guidelines exist for the design of outflow structures. In practice, designs vary considerably. However, silt traps are usually longer than they are wide by a factor of up to five. Silt trap depths vary considerably but are usually about 40 cm. However, it is often the lack of maintenance concerning outflow structures like silt traps that leads to problems years after the structure has been commissioned. Accumulated sediment has to be removed from silt traps to retain their original design performance and to avoid resuspension of accumulated pollutants during storms, causing pollution in the receiving water bodies (Pontier et al. 2001).

Watercourses in urban (built-up) areas that receive surface water runoff may be subject to pollution. This is particularly the case if silt traps, for example, are subject to insufficient maintenance (e.g., low frequency of sediment removal). Furthermore, the additional water load may contribute to local flooding (Pontier et al. 2001).

References

Convention on Wetlands of International Importance Especially as Waterfowl Habitat (1971) Wetland definition. Ramsar, Iran

Pontier H, Williams JB, May E (2001) Metals in combined conventional and vegetated road runoff control systems. Wat Sci Technol 44:607–614

Schmitt F, Milisic V, Bertrand-Krajewski JL, Laplace D, Chebbo G (1999) Numerical modeling of bed load sediment traps in sewer systems by density currents. Wat Sci Technol 39: 153–160

Scholz M (2006) Wetland systems to control urban runoff. Elsevier, Amsterdam, The Netherlands

Smith RL (1980) Ecology and field biology, 3rd edn. Harper and Row, New York, USA

Chapter 2
Wetland Case Studies

Abstract This chapter comprises six sections providing an overview of various representative and timely wetland case studies. The focus is on the treatment of domestic wastewater, farmyard runoff, swine wastewater, wood storage site runoff, and produced water containing hydrocarbons. Section 2.1 covers integrated constructed wetlands (ICW) for treating domestic wastewater. Section 2.2 provides guidelines for farmyard runoff treatment with farm constructed wetlands (FCW), a sub-type of ICW. Moreover, Section 2.3 covers a representative FCW case study on farmyard runoff treatment, and Section 2.4 describes ICW treating swine wastewater. Section 2.5 reviews wetlands controlling runoff from wood storage sites. Finally, Section 2.6 comprises a study reporting on wetlands for the treatment of hydrocarbons. All sections look at wetland systems from a holistic perspective, highlighting recent research findings relevant for practitioners. This chapter does not cover wetland systems such as sustainable (urban) drainage systems and sustainable flood retention basins used predominantly for flood control purposes. These techniques are discussed in detail in Chapter 4.

2.1 Integrated Constructed Wetlands for Treating Domestic Wastewater

2.1.1 Introduction

Sustainable wastewater treatment is associated with low energy consumption, low capital cost, and, in some situations, low mechanical technology requirements. Therefore, wetland treatment systems could be efficient alternatives to conventional treatment systems, especially for small communities, typically rural or suburban areas, due to low treatment and maintenance costs (Scholz *et al.* 2002; Soukup *et al.* 1994; Solano *et al.* 2003; Babatunde *et al.* 2008). Since the 1990s, wetland systems have been used for treating numerous domestic and industrial waste streams including those from the tannery and textile industries, slaughterhouses, pulp and paper production, agriculture (animal farms and fish farm effluents), and various runoff waters (agriculture, airports, highways, and storm water) (Kadlec et al. 2000; Haberl et al. 2003; Scholz 2006a; Vymazal 2007; Carty et al. 2008).

The concept of constructed wetlands applied to the purification of various wastewaters has received growing interest and is gaining popularity as a cost-effective wastewater management option in both developed and developing countries. Most of these systems are easy to operate, require low maintenance, and have low capital investment costs (Machate et al. 1997).

The treatment efficiency of most constructed wetlands depends on the water table level and the dissolved organic concentration of the influent (Reddy and D'Angelo 1997). The water level within most wetland systems (except for tidal vertical-flow constructed wetlands (Scholz 2006a)) is permanently kept above the wetland soils to create fully saturated soil conditions, generally resulting in high contaminant removal efficiencies. The treatment efficiencies of wetlands varies depending on the wetland design, type of wetland system, climate, vegetation, and microbial communities (Vacca et al. 2005; Ström and Christensen 2007; Picek et al. 2007; Weishampel et al. 2009).

The ICW concept was developed not only to address water pollution from different sources including domestic, industrial, and agricultural, but also to provide ecological services by restoring potentially lost environmental infrastructure including wetlands. The main features of ICW are shallow water depth, emergent vegetation, and the use of in situ soils that imitate those found in natural wetland ecosystems. No artificial liners (e.g., plastic or concrete) are used in the construction of ICW. Scholz et al. (2007a, b) and Babatunde et al. (2008) described the detailed concept and removal processes of these robust, sustainable, and synergistic systems by elucidating case studies in Ireland. Wastewater treatment in ICW systems takes place through various physical, chemical, and biological processes involving plants, microorganisms, water, soil, and sunlight (Kadlec and Knight 1996; Scholz 2006a; Mitsch and Gosselink 2007).

Although constructed wetlands are mechanically simple treatment systems, the passive treatment processes that remove contaminants are intricate; for instance, the hydrology, microbiology, and water chemistry are complex and interconnected. Research conducted on these systems demonstrates high removal percentages for 5-d biochemical oxygen demand (BOD), chemical oxygen demand (COD), suspended solids (SS), and pathogens, whereas nutrient removal percentages are usually low and variable. Mitsch and Gosselink (2007) claimed that effective nutrient removal could be achieved after a few growing seasons because of the lack of well-developed below- and aboveground plant–microbial interactions during the initial seasons. It is a common notion that the nutrient removal efficiency of constructed treatment wetlands decreases with age, especially for phosphorus removal as the mineral sediment becomes fully saturated; i.e., no free adsorption sites remain (Kadlec 1999).

Most previous studies were based on either pilot plant-scale or laboratory-scale experimental systems. Very few studies have been carried out on the assessment of performance of full-scale constructed wetlands treating domestic wastewater. There is currently no information on the performance of new and mature ICW systems treating domestic wastewater in the public domain. The purpose of this study was thus:

- to assess for the first time the treatment performance of an ICW system treating domestic wastewater on an industrial scale after 1 year of operation;
- to compare for the first time the annual and seasonal treatment efficiency of a full-scale mature and new ICW system; and
- to investigate the impacts of potential contamination of nearby surface waters and groundwater, taking into consideration that an artificial liner is not present.

2.1.2 Materials and Methods

2.1.2.1 Site Description

The case study systems comprise two ICW for treating domestic wastewater in Ireland. The ICW in Glaslough (Figures 2.1 and 2.2) is situated in the County of Monaghan (northern part of the Republic of Ireland), at a longitude of 06°53′37.94″W and a latitude of 54°19′6.01″N. The typical annual rainfall is approx. 970 mm over the last 50 years. However, the mean annual rainfall of 1256 mm was exceptionally high in 2008.

The system was commissioned in October 2007. Its purpose is to treat sewage and to contribute to the improvement of the water quality of the Mountain Water River. The inflow rate ranges between approx. 85 and 105 m^3/d. The corresponding outflow (approx. between 1 and 50 m^3/d) was very low due to evapotranspiration and infiltration of treated wastewater. The dilution of the wastewater due to rainfall on the wetlands is roughly between 35 and 65%, depending very much on the season and daily flow fluctuations.

The Glaslough ICW system has a design capacity equivalent to 1750 inhabitants and covers a total area of 6.74 ha. The water surface area of the constructed cells is 3.25 ha.

The ICW in Glaslough (Figure 2.1) consists of a small pumping station, two sludge cells, and five shallow vegetated cells. Domestic sewage from the village is pumped to the pumping station on site and from there to one of the sludge cells. There are two sludge collection cells that can be operated alternately to allow for subsequent desludging of the other cell if it is not in operation. From the sludge cell the wastewater flows by gravity through the five vegetated cells and the effluent finally discharges directly to the adjacent Mountain Water River. The wetlands were planted with *Carex riparia*, Curtis, *Phragmites australis*, (Cav.) Trin. ex Steud., *Typha latifolia*, L., *Iris pseudacorus* L., *Glyceria maxima* (Hartm.)

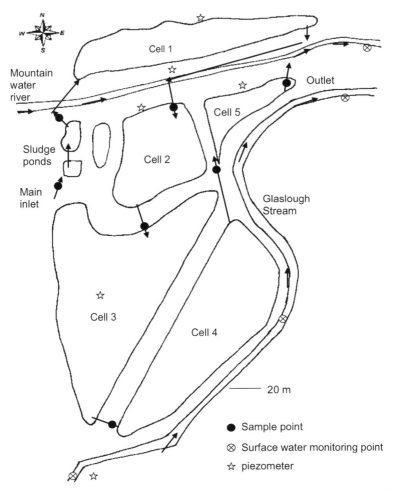

Figure 2.1 Sketch showing groundwater and surface water monitoring and inlet and outlet points for the integrated constructed wetland in Glaslough, near Monaghan, Ireland

Holmb., *Glyceria fluitans* (L.) R. Br., *Juncus effusus* L., *Sparganium erectum* L. emend Rchb, *Elisma natans* (L.) Raf. and *Scirpus pendulus,* Muhl. The ICW is flanked by the Mountain Water River and the Glaslough Stream.

Figure 2.2 Wetland cell in Glaslough, near Monaghan, Ireland

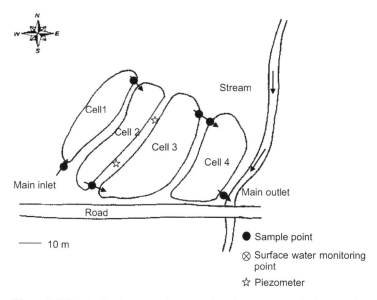

Figure 2.3 Sketch showing groundwater and surface water monitoring and inlet and outlet points for the integrated constructed wetland system in Dunhill, near Waterford, Ireland

The ICW system in Dunhill (County Waterford, southeastern part of Ireland) is situated at a longitude of 07°02'40"W and a latitude of 52°11'28"N (Figures 2.3 and 2.4). The typical annual rainfall is approx. 1000 mm. Until 2000, sewage at Dunhill village was directed to a wastewater treatment plant (septic tank system).

Figure 2.4 Integrated constructed wetland system in Dunhill, near Waterford, Ireland

In late 2000, the system was upgraded with the help of an ICW system that was fully operational by February 2001. The wastewater inflow was approx. $40 \, \text{m}^3/\text{d}$. The corresponding outflow was roughly $24 \, \text{m}^3/\text{d}$. Dilution of the wastewater due to rainfall is approx. between 5 and 20%, depending on season. However, detailed daily flow values are not available for this complex open system.

The main purpose was to treat sewage and to contribute to the improvement of the water quality of the Annestown stream. The system has a total area of 0.3 ha. The primary vegetation types used in the ICW are emergent plant species (helophytes). The system is gravity-fed and therefore has no energy consumption. Wastewater from households is collected via the sewage system and then transported to the wetland system. A single influent entry point is located in the first cell. The ICW system was based on four cells operating in series. The final effluent enters the Annestown stream via the outlet of the final ICW cell.

2.1.2.2 Sampling and Analytical Methods

Grab samples for the inlet and outlet of each wetland cell were taken approx. quarterly at the ICW in Dunhill, while a substantial suite of hi-tech automatic sampling and monitoring instrumentation has been used for approx. weekly sampling at the ICW in Glaslough: ISCO 4700 Refrigerated Automatic Wastewater Sampler (Teledyne Isco, Nebraska, USA), Siemens Electromagnetic Flow Meter F M MAGFLO and MAG5000 (Siemens Flow Instruments A/S, Nordborgrej, Nordborg, Denmark). Furthermore, both the flow rates (Figure 2.5) and key weather parameters (Figure 2.6) were automatically recorded.

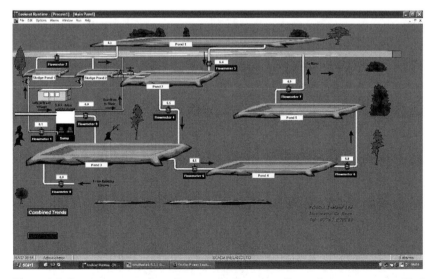

Figure 2.5 Computer screenshot of automatic flow recording software output in Glaslough, Ireland

Furthermore, the Mountain Water River and Glaslough Stream were also monitored (Figure 2.1). Water samples were analyzed for variables including the BOD, COD, SS, pH, ammonia nitrogen, nitrate–nitrogen, and molybdate reactive phosphorus (MRP; equivalent to SRP) at the Monaghan County Council water laboratory using American Public Health Association standard methods (APHA 1998) unless stated otherwise.

All cells consist of one inflow and one outflow structure, and the flow between each cell has been controlled by gravity through PVC pipes. Artificial liners were not used for both wetlands. However, the subsoil was worked and used as a natural liner.

Six piezometric groundwater-monitoring wells (Figure 2.1) were sampled at the Glaslough site to monitor the groundwater quality. The wells were placed within the ICW system and along the suspected flow path of contaminants towards the receiving watercourse. The ICW system was constructed using *in situ* soils. Subsoil obtained from the ICW site was reworked to line the ICW banks and cell beds to reduce groundwater infiltration and, subsequently, pollution. When polluted water flows through the ICW system, suspended solids settle naturally on the soil surface, obstructing infiltration of pollutants through the wetland cells (Kadlec and Knight 1996; Scholz 2006a; Wallace and Knight 2006).

The water table at the ICW site is relatively high (*i.e.*, 1.8 to 2.0 m below the ICW beds), so it is very important to monitor the ICW system and ensure that it has no negative effect on the groundwater. Six piezometers were placed at various depths (between 2.49 and 3.87 m). A site investigation by the Geological Survey of Ireland (IGSL Ltd., Unit F, M7 Business Park, Naas, County Kildare, Ireland) in September 2005 indicated a soil coefficient of permeability of approx.

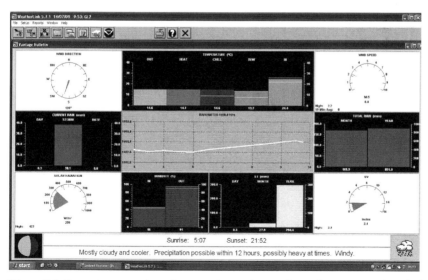

Figure 2.6 Computer screenshot of automatic weather station software output in Glaslough, Ireland

9×10^{-11} m/s. Piezometer 1 near wetland cell 1 (close to a small hill) and piezo-meter 6 (located across the Glaslough stream) are outside of the ICW system (Figure 2.1).

For the ICW system in Dunhill, two piezometric groundwater-monitoring wells were sampled at a depth of 5 m (Figure 2.3). The wells were placed within the ICW system and along the suspected flow path of contaminants to assess the risk of groundwater contamination. The subsequent water quality analysis for both ICW systems was carried out according to APHA (1998).

The stream water quality adjacent to the ICW systems was regularly monitored to assess the impact of ICW on receiving waters and verify that the ICW discharge was not polluting the receiving waters. The Mountain Water River is sampled at two locations, one upstream and one approx. 400 m downstream of the discharge point. Moreover, the Glaslough stream (not a directly receiving watercourse) is monitored at three points.

The Annestown stream near the ICW in Dunhill is also monitored to check for compliance with discharge standards. The two sampling points are located approx. 4 km upstream and 3.5 km downstream of Dunhill village.

2.1.2.3 Statistical Analyses

All statistical analyses were carried out using the computer software package Origin 7.5. A parametric analysis of variance (ANOVA) was used to determine any significant ($p < 0.05$) differences in removal percentages and the seasonal effect on water quality for both ICW systems.

2.1.3   Results and Discussion

2.1.3.1   Water Quality of the ICW System in Glaslough

The approximate mean inflow values in Glaslough were as follows: BOD, 768 ± 451.0 mg/l; COD, 1279 ± 697.8 mg/l; SS, 2184 ± 3844.8 mg/l; ammonia–nitrogen, 32 ± 11.1 mg/l; nitrate–nitrogen, 5 ± 3.8 mg/l; MRP, 4 ± 2.0 mg/l; pH, 7 ± 0.4. These values indicate a very high variability of the domestic wastewater entering the ICW system (Kayranli *et al.* 2010).

However, the ICW system has shown a very good treatment performance, de-spite being a new (*i.e.*, not mature) system. The ICW system removed approx. 99% of BOD, 97% of COD, 100% of SS, 99% of ammonia–nitrogen, 94% of nitrate–nitrogen, and 99% of MRP during this period. The results show that the pollutant removal capacity is very high due to its large wetland size, providing high mean retention times. Findings contrast with the general idea that the organic matter removal rate increases depending on the constructed wetland's age (Kadlec 1999). An increase in age is also associated with an increase in microbial population. Furthermore, microbial biofilm formation on the bed material within the wetland leads to higher biological degradation rates (Picard *et al.* 2005).

There is very little information on full-scale representative constructed wetlands for treating domestic wastewater in the scientific literature. However, numerous studies refer to pilot-scale and microcosm-scale constructed wetlands for treating domestic wastewater. Ciria *et al.* (2005) assessed the role of *Typha latifolia* L. (reedmace, cattail, or bulrush) in constructed wetlands of 40 m^2 each, filled with gravel as the supporting medium. Their study showed that BOD and COD removal efficiencies were $97 \pm 1.2\%$ and $79 \pm 0.3\%$ in the first year, respectively, and in the second year, the BOD removal efficiency did not change ($97 \pm 3.0\%$), while the COD removal efficiency increased slightly ($81 \pm 1.0\%$). Furthermore, a study by Hamouri *et al.* (2007) achieved removal efficiencies of 78% for COD and 79% for BOD with respect to the combination of a two-step up-flow anaerobic reactor and sub-surface horizontal-flow constructed wetland (Hamouri *et al.* 2007). Between 71 and 75% removal efficiencies for COD and BOD were noted for constructed wetlands treating secondary treated sewage (Thomas *et al.* 1995).

The main nitrogen removal process within constructed wetland systems include uptake from plants and other living organisms, sedimentation, nitrification, denitrification, ammonia volatilization, and cation exchange for ammonium (Newman *et al.* 1999; Yang *et al.* 2001; Scholz 2006a; Wallace and Knight 2006; Mitsch and Gosselink 2007). Kadlec *et al.* (2000) have explained that interactions between nitrogen on one hand and water, sediment, plant, and biomass on the other make it difficult to assess the real efficiency of nitrogen removal due to storage in the system.

The overall ammonia–nitrogen reduction is relatively high in comparison to other microcosm wetlands treating domestic wastewater: 76 to 92% for sub-surface constructed mangroves in Hong Kong (Wu *et al.* 2008); 52% for a combination of free surface-flow wetland cells that were fed with municipal lagoon effluents in Canada (Cameron *et al.* 2003); 10 to 20% for a continuous-flow, free water surface pilot wetland planted with *Lemna gibba* L. (duck weed) in Israel (Ran *et al.* 2004); 9% for a sub-surface horizontal-flow constructed wetland in Morocco (Hamouri *et al.* 2007); and 14 to 24% for a constructed wetland treating secondary treated sewage in Australia (Thomas *et al.* 1995).

Seasonal variations in performance related to nutrient removal were also investigated. The BOD concentrations within the influent and the effluent were relatively high in summer and autumn compared to other seasons. However, the BOD removal efficiency in summer was similar to those of the other seasons. The COD concentration in the influent was lower in spring compared to the other seasons. However, lower COD concentrations within the effluent (27.0 ± 9.52 mg/l; removal efficiency of 98%) were recorded during winter in comparison with concentrations (52.6 ± 53.48 mg/l; removal efficiency 96%) in spring. In contrast, the lower effluent concentrations (1.3 ± 1.49 mg/l) were recorded for SS in autumn. Hunt and Poach (2001) explain that constructed wetland systems cannot completely remove carbon and solid compounds because wetland plants produce plant litter, which continuously adds carbon and other compounds to the system.

The highest ammonia–nitrogen and nitrate–nitrogen concentrations within the influent were 35.3 ± 9.87 mg/l in spring and 7.7 ± 2.60 mg/l in summer, respectively. On the other hand, the lowest ammonia–nitrogen and nitrate–nitrogen concentrations in the effluent were 0.0 ± 0.15 mg/l in autumn (removal efficiency of 100%) and 0.2 ± 0.12 mg/l (removal efficiency of 83%) in winter, respectively. The effluent ammonia–nitrogen concentrations were slightly higher in winter compared to the other seasons. This can be explained by the fact that nitrification of ammonia–nitrogen is relatively low in winter due to the lack of oxygen and low temperatures, which negatively affect nitrification rates within constructed wetlands (Kayranli et al. 2010). Moreover, the effluent nitrate–nitrogen concentration was slightly different irrespective of seasonal change, although the influent nitrate–nitrogen concentration was considerably higher in summer and autumn than in spring and winter. This indicates that the denitrification rate for the system was rather high, especially in summer and autumn due to high temperatures, elevated microbial activity, and the presence and easy availability of organic carbon.

Oxygen is generally used for nitrification and organic matter reduction. However, oxygen is partly generated due to photosynthesis during the daytime and supports the oxygen demand for stabilization of organics and nitrification. Plant rhizosphere aeration may stimulate aerobic decomposition processes, increasing nitrification and subsequent gaseous losses of nitrogen via denitrification and decreasing relative levels of dissimilatory nitrate reduction to ammonium (Tanner et al. 1995). Microbial activity and plant nutrient uptake within wetlands are both directly and indirectly affected by temperature. Nitrification is a temperature-dependent process. Werker et al. (2002) reported that nitrification rates in constructed wetlands become increasingly inhibited at temperatures of about 10°C, and rates drop rapidly at 6°C. Akratos and Tsihrintzis (2007) noted high temperature requirements for the removal of nitrogen compounds such as ammonia at temperatures above 15°C. Temperature sharply affected the system performance due to changing biological activities with temperature.

Ciria et al. (2005) reported that ammonia–nitrogen and nitrate–nitrogen reduction rates were significantly higher in the second compared to the first year of operation, and they obtained the highest removal rates of $22 \pm 0.8\%$ for ammonia–nitrogen and $64 \pm 0.9\%$ for nitrate–nitrogen in autumn of the first year and $40 \pm 3.5\%$ for ammonia–nitrogen in summer and $75 \pm 2.1\%$ for nitrate–nitrogen in winter for the second year. In comparison to the previous study, the ammonia–nitrogen and nitrate–nitrogen removal efficiencies for all seasons are significantly higher for the ICW system.

Phosphate removal mechanisms are based on physical (sedimentation), chemical (adsorption), and biological processes. Phosphorus can be reduced directly by plant uptake or chemical storage within sediments (Bonomo et al. 1997). In contrast, Sakadevan and Bavor (1998) suggested that long-term phosphorus removal mechanisms in constructed wetland systems are likely due to uptake by the substratum, litter, and aluminum/iron compounds, while plant uptake is often a relatively small fraction. Furthermore, Kadlec (1999) reported that the phosphorus reduction capacity decreases with age because the mineral sediments become fully saturated within the wetland systems, *i.e.*, no free adsorption sites remain. The annual MRP reduction rate for Glaslough was 99.2%, which means that the system has a high phosphorus adsorption and storage capacity and high plant uptake capacity. However, the system is relatively young as well.

2.1.3.2 Receiving Stream Water Quality

The Mountain Water River is adjacent to the sludge cells and to wetland cells 1, 2, and 5, whereas the Glaslough Stream is adjacent to wetland cells 4 and 5. At the discharge point, mean effluent concentrations were 0.4 mg/l for ammonia–nitrogen, 1.0 mg/l for nitrate–nitrogen, and 0.1 mg/l for MRP, whereas mean river downstream concentrations were 0.40 mg/l for ammonia–nitrogen, 1.02 mg/l for nitrate–nitrogen, and 0.1 mg/l for MRP. The river water quality did not change in the downstream compared to the upstream stretch. This indicates that the river has sufficient assimilative capacity.

In summer, the SS and nitrate–nitrogen concentrations in the Mountain Water River were slightly higher downstream than upstream, which was, however, not statistically significant ($p < 0.05$). Similarly, the BOD and COD concentrations of the influent and effluent were higher in summer than during the other seasons. This can be explained by the decrease in flow rate, and also by high evaporation rates in summer, while pollutant loads remain similar. The river and stream water quality is variable and depends predominantly on precipitation patterns. Furthermore, especially in spring and autumn, high precipitation rates occur during periods when the buffering capacity of the receiving water is enhanced by an increased dilution ratio.

The Mountain Water River had mean ammonia–nitrogen, nitrate–nitrogen, and MRP concentrations of 0.4 ± 0.28 mg/l, 1.0 ± 0.41 mg/l, and 0.1 ± 0.06 mg/l for the upstream and 0.4 ± 0.28 mg/l, 1.0 ± 0.38 mg/l, and 0.1 ± 0.06 mg/l for the downstream stretches, respectively. The Glaslough Stream, which originally passed through the site, was diverted and widened around the perimeter of cell 4. There is no discharge point from the ICW system into the Glaslough Stream, but there are three sampling points to assess potential contamination from the ICW. Concerning the mean water quality values, ammonia–nitrogen, nitrate–nitrogen, and MRP concentrations were 0.3 ± 0.32 mg/l, 1.2 ± 0.37 mg/l, and 0.1 ± 0.07 mg/l, respectively. The water quality was better downstream than upstream. The ammonia–nitrogen, nitrate–nitrogen, and MRP concentrations for the sample points adjacent to cell 5 and after cell 5 were 0.3 ± 0.12 and 0.2 ± 0.11 mg/l, 0.9 ± 0.44 and 0.8 ± 0.50 mg/l, and 0.1 ± 0.07 and 0.2 ± 0.13 mg/l, respectively. This indicates that the ICW system does not pollute the nearby Glaslough Stream.

Molybdate reactive phosphorus effluent concentrations are in compliance with the Irish Phosphorus Regulations (1998) (Environmental Protection Agency 1998), which set an annual median threshold of 0.03 mg/l for rivers. However, neither the Mountain Water River nor the Glaslough Stream upstream of the ICW system complied with this regulation.

Concerning the Annestown Stream, the MRP concentration at the monitoring station Ballyphilip (4 km upstream of the ICW system) was well below the target phosphorus concentration of 0.03 mg/l, set by the Irish Phosphorus Regulations from 1998 (Environmental Protection Agency 1998). However, there has been a very slight increase (0.002 mg/l) in the MRP concentration 3.5 km downstream of the ICW system.

A slight increase in nutrient concentrations between the two monitoring points upstream and downstream of the ICW system was noted. However, only ammonia–nitrogen increased significantly ($p < 0.05$), while increases in nitrate–nitrogen and MRP concentrations were statistically insignificant. The increase in nutrient concentration downstream of the ICW system may be attributed to runoff containing nutrients originating from intensive cattle farming. Furthermore, the ICW system is overloaded; over the course of time, new housing developments in Dunhill have led to an increase in sewage.

2.1.3.3 Groundwater Quality

The mean ammonia–nitrogen, nitrate–nitrogen and MRP concentrations within piezometer 1 were 0.5 ± 0.28, 0.3 ± 0.15, and 0.2 ± 0.41 for MRP, respectively. Piezometer 2 is located in the west of the ICW system near cell 1, whereas piezometers 3 and 4 can be found in the east near cells 2 and 5. The mean ammonia–nitrogen, nitrate–nitrogen, and MRP concentrations for piezometer 2 were 0.7 ± 0.42 mg/l, 0.2 ± 0.10 mg/l, and 0.4 ± 0.31 mg/l, which indicate a slight increase in concentrations. No groundwater contamination was observed. The water quality characteristics of piezometers 3 and 4 concerning ammonia–nitrogen, nitrate–nitrogen, and MRP were as follows: 4.6 ± 4.42 mg/l and 0.3 ± 0.28 mg/l, 0.8 ± 0.51 mg/l and 0.5 ± 0.55 mg/l, and 0.4 ± 0.53 mg/l and 0.2 ± 0.50 mg/l, respectively. The mean water quality concentrations for piezometer 3 are higher than for piezometer 4. This can be explained by the observation that piezometer 3 is located near wetland cell 2, which contains more pollutants than other cells (except for cell 1). Furthermore, low infiltration takes place within cell 2 due to the presence of a sandy layer at the bottom. The mean water quality values regarding piezometer 6, which is located across the Glaslough Stream, were 2.8 ± 1.51 mg/l, 0.9 ± 0.83 mg/l, and 0.2 ± 0.20 mg/l for ammonia–nitrogen, nitrate–nitrogen, and MRP, respectively. The ammonia–nitrogen concentrations were often high at piezometer 6, which is located near an equestrian center, potentially polluting the groundwater.

In Dunhill, piezometer 1 near wetland cell 2 and piezometer 2 near cell 3 are located within the ICW system to monitor groundwater quality (see also Figure 2.3). Concerning piezometer 1, 4.7 ± 12.25 mg/l and 0.4 ± 0.93 mg/l have been measured for ammonia–nitrogen and MRP, respectively. The water quality for piezometer 2 was as follows: ammonia–nitrogen and MRP had mean concentrations of 2.6 ± 1.16 mg/l and 0.0 ± 0.02 mg/l, respectively. This can be explained by the fact that the ammonia–nitrogen reduction due to the ICW system is relatively low because of overloading. Moreover, some infiltration may occur from cell 2. Concerning MRP, values were low within the piezometer sample water. It is most likely that MRP is taken up by the soil via adsorption (Kayranli *et al.* 2010).

2.1.3.4 Comparison of Nutrient Removal Performances

Overall, both ICW systems indicate significant ($p < 0.05$) COD and BOD removal efficiencies. Concerning the other water quality data, ammonia–nitrogen, nitrate–nitrogen, and MRP removal efficiency for the Glaslough system were high: 99.0%, 93.5%, and 99.2%, respectively. In comparison, the ICW in Dunhill had removal efficiencies of 58%, −80.8% (source rather than sink), and 34.0%, respectively. Nitrate–nitrogen and MRP concentrations within the effluent gradually increased. The decreasing nutrient removal rates for Dunhill are probably due to the increased system overload. Ammonia–nitrogen, nitrate–nitrogen, and MRP concentrations within the effluent are three times higher for the fourth year of its operation than for the previous three years. Nitrate–nitrogen concentrations within the effluent were higher than for the influent, which means that some ammonia–nitrogen is transferred into nitrate–nitrogen via nitrification. However, ammonia–nitrogen and nitrate–nitrogen were both released from the ICW in Dunhill.

The organic material present within the ICW cells also has an indirect impact on the bacterial community. For instance, the litter on top of the sediment is likely to have limited the diffusion of oxygen to the lower sediment layers, creating anoxic conditions and, hence, making conditions favorable for denitrification. This possible process has been described previously by Bastviken *et al.* (2005) for a comparable system. Most denitrifiers are heterotrophs, and the supply of organic carbon by macrophytes may have raised the overall heterotrophic activity, leading to the consumption of oxygen (Souza *et al.* 2008). Thus, it is likely that the oxygen availability within the sediment was reduced, and denitrification was subsequently supported (Bastviken *et al.* 2005).

The COD and BOD effluent concentrations within both ICW systems were generally higher in summer and autumn than in winter and spring. However, the removal efficiencies for these parameters did not change significantly for either system. This is probably due to the increase in organic loading rate as a consequence of an increase in the evaporation rate and a decrease in the precipitation rate. On the other hand, effluent ammonia–nitrogen, nitrate–nitrogen, and MRP concentrations did not change significantly for the ICW system in Glaslough, whereas these variables increased for the system in Dunhill.

The difference in removal rate could also be due to higher hydraulic retention times provided for the ICW system in Glaslough, performing better in terms of MRP removal compared to the system in Dunhill. The Glaslough system removed 99.2% more MRP than the ICW system in Dunhill. The difference in MRP reduction is likely due to Glaslough's subsoil and sediment, which may not have reached the saturation threshold (Kayranli *et al.* 2010).

2.1.3.5   Comparison of Nutrient Reduction in Wetland Cells

The ICW systems operate as sequential multicellular structures and have a minimum number of four wetland cells. The influent (*i.e.*, effluent from the sedimentation cell) to the first wetland cell in Glaslough has the following characteristics: BOD, 405.1 ± 204.03 mg/l; COD, 773.0 ± 506.67 mg/l; SS, 238.3 ± 158.03 mg/l; ammonia–nitrogen, 37.3 ± 10.73 mg/l; nitrate–nitrogen, 3.6 ± 2.54 mg/l; MRP, 4.1 ± 1.89 mg/l. The corresponding effluent of cell 1 is as follows: 35.6 ± 29.59 mg/l for BOD, 127.3 ± 71.00 mg/l for COD, 30.6 ± 44.06 mg/l for SS, 18.4 ± 7.53 mg/l for ammonia–nitrogen, 1.3 ± 1.13 mg/l for nitrate–nitrogen, and 3.2 ± 1.24 mg/l for MRP.

In comparison, the influent (*i.e.*, effluent from a septic tank) to the ICW in Dunhill has the following water quality characteristics: 358.4 ± 200.57 mg/l for BOD; 554.4 ± 288.19 mg/l for COD; 303.5 ± 335.46 mg/l for SS; 52.6 ± 39.30 mg/l for ammonia–nitrogen; 0.6 ± 1.74 mg/l for nitrate–nitrogen; 7.8 ± 3.38 mg/l for MRP. An improvement due to treatment in wetland cell 1 was noticed: 55.3 ± 28.65 mg/l for BOD; 149.6 ± 69.80 mg/l for COD; 42.8 ± 36.71 mg/l for SS; 49.9 ± 37.92 mg/l for ammonia–nitrogen; 1.1 ± 1.37 mg/l for nitrate–nitrogen; 7.0 ± 4.69 mg/l for MRP.

These findings indicate that variables including BOD, COD, and SS were significantly reduced within the first cell of both systems even after 5 years of ICW operation in Dunhill. Nitrate–nitrogen and MRP concentrations were reduced significantly within the first cell, whereas the ammonia–nitrogen reduction rate was higher in wetland cell 2 than in cell 1. However, the ICW in Glaslough had a higher pollutant reduction capacity than that in Dunhill. This is most likely due to overloading of the ICW system in Dunhill.

As can be seen from Figures 2.7–2.9, ammonia–nitrogen and MRP were removed in significant ($p < 0.05$) amounts after the contaminated water passed approx. 30% of the ICW area in Glaslough, whereas the COD, BOD, and SS were reduced after passing 20% of the ICW area. This can be explained by the low COD/BOD ratio of 1.45, which means that most pollutants in the contaminated water are biodegradable. Concerning the ICW in Dunhill, nutrient reduction rates were low in the first cell (Figures 2.10 and 2.11). The BOD, COD, and SS concentrations were mostly reduced within the first cell (Figure 2.12).

Kadlec *et al.* (2000) and Carty *et al.* (2008) report that nutrient reductions occur predominantly within the initial wetland cells (as confirmed in this study, except for MRP) and that pollutants are reduced effectively, if the hydraulic retention time is relatively high. This is promoted by allowing the pollutant plume to spread as slowly as possible throughout flatly designed ICW cells.

The nitrate–nitrogen concentrations within the ICW systems were low. It is likely that nitrate and oxygen provided electron acceptors in the lower layer of the wetland cells (Eriksson and Weisner 1996). Furthermore, Nielsen *et al.* (1990) reported that the high number of denitrifying bacteria was dependent on the accumulation of plant detritus within ICW systems. The overall heterotrophic activity is increased by the supply of sufficient organic matter. This leads to the consumption and subsequent reduction of oxygen within the sediment, thus supporting denitrification.

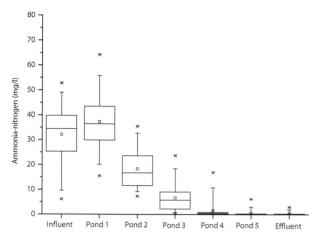

Figure 2.7 Ammonia–nitrogen concentrations for the integrated constructed wetland system in Glaslough, Ireland

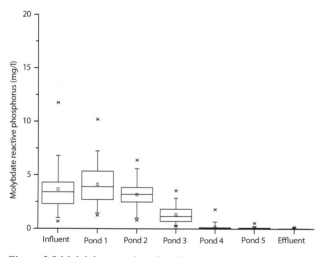

Figure 2.8 Molybdate reactive phosphorus concentrations for the integrated constructed wetland system in Glaslough, Ireland

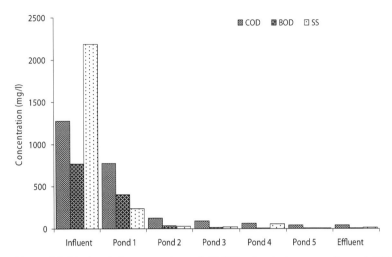

Figure 2.9 Biochemical oxygen demand (BOD), chemical oxygen demand (COD), and suspended solid (SS) concentrations for the integrated constructed wetland system in Glaslough, Ireland

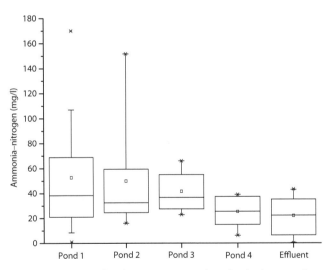

Figure 2.10 Ammonia–nitrogen concentrations for the integrated constructed wetland system in Dunhill, Ireland

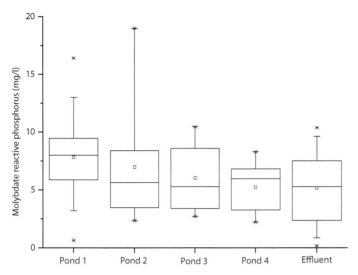

Figure 2.11 Molybdate reactive phosphorus concentrations for the integrated constructed wet-land system in Dunhill, Ireland

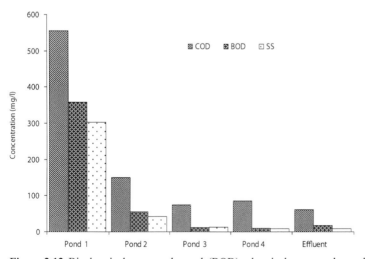

Figure 2.12 Biochemical oxygen demand (BOD), chemical oxygen demand (COD), and suspended solid (SS) concentrations for the integrated constructed wetland system in Dunhill, Ireland

2.1.4 Conclusions and Further Research Needs

The ICW concept has been successfully applied for the first time on an industrial scale for the treatment of domestic wastewater in real case studies that were fully scientifically monitored and assessed. The new and mature ICW systems successfully removed traditional pollutants such as BOD from domestic wastewater. Concerning the new ICW system (1 year of operation) in Glaslough, the nutrient reduction efficiencies are significantly high, whereas the nutrient reduction efficiencies (including nitrate–nitrogen and MRP) started to decrease after 5 years of operation due to overloading of the mature ICW. However, the BOD, COD, SS, and ammonium–nitrogen concentrations were reduced within the mature ICW system even after approx. 5 years of operation. However, while nitrification of ammonia–nitrogen was significant, the denitrification rate started to decrease as the ICW matured.

Both groundwater and surface water monitoring results indicated that the ICW system in Glaslough had neither polluted the groundwater nor decreased the water quality of the receiving watercourse. All nutrient concentrations for the receiving watercourse were lower downstream than upstream of the ICW system outlet. On the other hand, nutrient removal within the open ICW system is complex due to water, sediment, plant, and microbial interactions, so it was impossible to come up with consistent nutrient balances.

The novel use of ICW to treat domestic wastewater is a valuable and appropriate technology. It is especially suitable for small communities in both developed and developing countries. The absence of an artificial liner made of materials such as plastic or concrete makes the ICW technology affordable. However, any ICW system should be mature and sufficiently large to avoid potential groundwater contamination.

There is scope for further research on the assessment of water balances and processes responsible for the self-sealing effect observed in mature wetlands. A detailed assessment of the microbial population dynamics and the role of species influencing treatment performance would therefore be beneficial.

2.2 Guidelines for Farmyard Runoff Treatment with Wetlands

2.2.1 Introduction

The loss of nutrients and contaminants from agricultural land, farmyards (Figure 2.13), dairy parlors, tracks, and roofs to rivers, lakes, and groundwater can have a detrimental impact on water quality. Both point and diffuse sources of pollution from agriculture contribute to the degradation of water quality and aquatic ecosystems (e.g., fish kills and loss of habitat) through eutrophication, contamination of groundwater, siltation, and direct toxicity to organisms, which

Figure 2.13 Typical farmyard in Scotland

consequently affect biodiversity, fisheries, recreation, and public health. They also affect farmers, exposing them to fines and prosecutions, and the wider community by the subsequent degradation and loss of water supply within affected watersheds (Harrington and Ryder 2002; Mantovi *et al.* 2003; Scholz *et al.* 2007a, b).

There is a wide range of laws and policies at national and international levels (*e.g.*, Zedler 2003 and Scholz 2006) that aim to control and mitigate the risks of water pollution caused by agricultural contaminants. Several BMP are available for farmyard runoff treatment as part of "treatment trains" (Scholz 2006a, b); *i.e.*, sets of measures that range from pollution source control to dirty water collection and treatment. The most common measures implemented at the farm scale include the following (Rice *et al.* 2002; Hilton 2003):

- animal diet improvement to reduce nutrient losses;
- roofing of silage pits and areas of farmyards where excrement is expected to accumulate;
- clean roof water diversion to drains to reduce the volume of dirty water to be stored and spread;
- upgrading of buildings and of manure, slurry, fuel, and pesticide storages to avoid leaks and spillages;
- basins and biobeds for pesticide wash water storage and treatment; and
- swales, buffer strips, and wetlands to store and clean farmyard runoff before discharge to watercourses (Poe *et al.* 2003).

In addition to the previously mentioned BMP, farm constructed wetlands (FCW) are recommended for the collection and treatment of farmyard runoff, before it is released into watercourses, to protect surface water and groundwater resources (Scholz *et al.* 2007a, b). The US and New Zealand approaches to FCW have been summarized by USEPA (1995a, b) and NIWA (1997), respectively.

2.2.2   Farm Constructed Wetlands: Definition and Background

2.2.2.1   Introduction

An FCW is defined as an ecologically engineered system comprising a series of shallow, free-surface-flow constructed wetland cells (see below) containing emergent vegetation, which is designed to receive and treat farmyard runoff (Carty *et al.* 2008; Scholz *et al.* 2007a, b). Most FCW are being developed to benefit both the environment and farmers by reducing the impact of potential pollution incidents, helping to manage farm effluents, and enhancing habitat, biodiversity, and landscape in a way that is practical, efficient, affordable, and cost-effective (Harrington *et al.* 2005). Typical FCW design and operation contrast with those of constructed reed beds, which can be seen as being 'closer' to the traditional wastewater treatment philosophy based predominantly on civil engineering principles (Mantovi *et al.* 2003; Kantawanichkul and Somprasert 2005; Sun *et al.* 2006).

The design approach for the construction of farm wetlands proposed in this universal guideline for temperate climate is largely inspired by the ICW concept (Harrington and Ryder 2002; Scholz *et al.* 2007a, b) pioneered in Ireland by the National Parks and Wildlife Service (Department Environment Heritage and Local Government, Ireland) and is based upon data on performance over the last 10 years of 13 ICW constructed within the Anne Valley in southern Ireland. The ICW concept is based on the following principles (Harrington *et al.* 2005):

- the containment and treatment of influents within shallow vegetated ponds using local soil material, wherever possible;
- the aesthetic placement of the containing wetland structure into the local landscape ("landscape fit") to enhance a site's ancillary values; and
- the enhancement of habitat and biodiversity.

The design endeavors to optimize natural biological, chemical, and physical processes of pollutant removal in a way that does not produce a negative impact on aquatic and terrestrial ecosystems (Scholz *et al.* 2007a, b).

2.2.2.2   Effluent Types and Processes

All FCW can be designed to treat various types of wastewater from farms including dairy, machinery, vegetable and mushroom washings, runoff from silos, yards and other areas of hard-standing (usually only lightly contaminated by manure or silage), and dairy cow access tracks (Poe *et al.* 2003; Carty *et al.* 2008; Scholz *et al.* 2007). However, FCW are usually not designed for the treatment of more nutrient-rich effluent types such as slurries, raw milk, and washings from pesticide sprayer and dipping equipment.

Within an FCW, the contaminated effluent is treated through various physical, chemical, and biological processes involving plants (predominantly macrophytes), algae, microorganisms, water, soil, and sun (*e.g.*, direct process via photodegradation). The main processes for which detailed examples can be found elsewhere (Carty *et al.* 2008; Mantovi *et al.* 2003; Scholz and Xu 2002; Scholz and Lee 2005; Scholz 2006a; Scholz *et al.* 2007a, b; Zedler 2003) are as follows:

- physical filtration of suspended solids by wetland vegetative biomass acting as a hydrological baffle to incoming flows (optimized by high vegetation density and low flow velocity);
- settling of suspended particulate matter by gravity (optimized by low flow velocity, low wind speed, low disturbance, and long residence time);
- uptake, transformation and breakdown of nutrients, hydrocarbons and pesticides by biomass, plants and microbes (increased by a relatively high temperature, long residence time, contact with microorganisms and plants, high microorganism and plant density, and a relatively high organic matter content);
- accumulation and decomposition of organic matter, which is important for nutrient cycling (optimized by low velocity and availability of adsorption sites on suitable aggregates);
- microbial-mediated processes such as nitrification (aerobic) and denitrification (anaerobic), which are important for the cycling of nitrogen (see also Poe *et al.* (2003));
- chemical precipitation and sorption of nutrients such as phosphorus (see also Braskerud (2002)) by soil (influenced by the availability of sorption sites, pH, and redox potential); and
- predation and natural dieoff of pathogens (*e.g.*, *Escherichia coli* and *Cryptosporidium*) (optimized by high diversity and density of natural predators (*e.g.*, protozoa), and increased exposure to ultraviolet light).

2.2.2.3 Functions, Values, and Principles

The profiles and infrastructural details required to support habitat development and biodiversity enhancement are, on the basis of experience, best addressed in the design and construction of the wetland. This is particularly relevant to the development of transitional habitats between the terrestrial embankment and wetland zones (Scholz and Lee 2005; Scholz 2006a).

Wide, shallow, and low elevated embankments promote floral and faunal diversity. Shallow and deep wetland areas with either south- or north-facing aspects are also important to enhance the habitat. Local vegetation is best incorporated wherever possible. Care should be taken when locating an FCW immediately adjacent to woodland as problems may arise from shading and seepage of water via root systems. Many FCW can be dynamic ecosystems and their habitats may change unless managed and maintained. For example, the management of water depth to facilitate optimal water treatment is particularly important in this regard (Dunne *et al.* 2005a, b; Scholz 2007; Scholz *et al.* 2007a, b; see also below).

Wetlands, including FCW, can have considerable aesthetic appeal. The combination of water, vegetation, and associated wildlife are the principal elements of visual enhancement, and there are many examples of this throughout the world. The aesthetic appeal of an FCW can be maximized through appropriate land-forming design that is implemented during construction (Scholz 2007).

The process of design ensures that the final wetland structure 'fits' well into the landscape, e.g., by designing curvilinear enclosing embankments that conform to a site's topography. Subsequent vegetation development will further enhance the visual natural appearance of the system. Appropriate land forming of the structure to fit the landscape also reduces FCW maintenance, thus enhancing a variety of amenity values and improving its sustained functionality (Dunne et al. 2005a, b; Scholz 2006b; Scholz et al. 2007a, b).

2.2.2.4 Benefits of Farm Constructed Wetlands

The main benefits of FCW as discussed by Zedler (2003), Scholz and Xu (2002), Scholz et al. (2007), and other researchers (see below) are summarized below:

- High level of treatment and robustness: efficient treatment (up to 99% reduction by concentration) of contaminants such as phosphorus (Braskerud 2002), nitrate–nitrogen, nitrite–nitrogen, ammonia–nitrogen, BOD, suspended solids, hydrocarbons, and pathogens.
- Runoff and flood management: FCW are designed sufficiently large to manage heavy rainfall events, providing attenuation for increased effluent volumes during storm events. By functioning as 'buffer zones' and attenuating peak flows, they contribute to the reduction of flooding incidents downstream (Scholz and Lee 2005; Scholz 2007).
- Relatively low cost and simplicity of operation: FCW have minimal equipment needs and little, if any, energy use since water can be transferred by gravity through the system. They are also simple to operate and more cost-effective than alternative methods of disposing of farm runoff (Poe et al. 2003).
- Odor minimization: odor can be a serious problem when handling and treating agricultural wastes. Odors are minimized in FCW through the use of a dense plant cover, appropriate plant species, a shallow water level, and surface-flow maintenance.
- Aesthetically pleasing: FCW enhance the landscape by adding colors and texture and by increasing the diversity of plants and habitats (Scholz 2007).
- Habitat and biodiversity enhancement: FCW provide habitats for a wide variety of birds, mammals, reptiles, amphibians, and invertebrates (e.g., Froneman et al. 2001).
- Contingency measures: FCW can help to mitigate the impact of accidental spillages, acting as buffer zones and giving time to implement emergency measures.

2.2.2.5 Limitations of Farm Constructed Wetlands

All FCW have some limitations, although these can be managed by designing and maintaining FCW for their design goals, as outlined in this sub-section. The main limitations as discussed by Scholz *et al.* (2007) and Carty *et al.* (2008) are as follows:

- Some FCW have a relatively large land requirement compared to conventional wastewater treatment systems (Scholz 2006).
- The removal of pollutants may vary during the year and in the long term due to seasonal weather patterns (*e.g.*, as reported by Dunne *et al.* (2005b)) and also variations in the inflow of pollutants. In some circumstances, pollutants might even be released (*e.g.*, as reported by Thorén *et al.* (2004)), but this can be minimized by the use of a modular approach (*i.e.*, using a series of FCW cells) and by designing the system for extreme rainfall events.
- Emission of greenhouse gases (*e.g.*, methane and nitrous oxide as reported by Hefting *et al.* (2003)).
- Most FCW should be regarded primarily as water treatment systems and treated accordingly; *i.e.*, not used for bathing, fishing, or animal watering due to the possible presence of pathogens, toxins, and parasites.
- Although safety and health concerns for humans and livestock may arise from FCW because they contain standing water, such concerns can be minimized by the use of gently sloping sides, marginal vegetation, and by raising public awareness.
- Some FCW pollutants (*e.g.*, heavy metals and pesticides) may cause harm to wildlife, but this impact can be mitigated by an appropriate design taking into account ecological aspects and by keeping certain types of effluents (*e.g.*, pesticide washings) away from the FCW.
- Some infiltration of water to groundwater may occur. However, infiltration is reduced by the use of shallow water depths and an adequate substrate such as clay (Hill *et al.* 2000) and decreases over time through sealing of the bed of the wetland by accumulated organic matter and sediment (Harrington *et al.* 2005).

2.2.3 Farm Constructed Wetland Site Suitability

2.2.3.1 Effluent to Be Treated

In order to decide if a FCW is needed and appropriate for a given farm, a site-specific approach is required, usually involving the following steps (Zedler 2003; Carty *et al.* 2008):

1. assessment of the type and volume of the effluent to be treated (present and future) and the infrastructure present on the farm;
2. determination of the need for a FCW from an assessment of existing on-farm measures and potential cost-effective alternatives for dirty water management; and

3. determination of the characteristics of the site available for FCW construction and assessment of any potential impacts that may result from FCW implementation.

It is recommended that specialist advice from a suitably qualified consultant or engineer be sought when assessing the need for a FCW and the suitability of an FCW site. To decide whether a FCW is an appropriate option for a farm, the farmer or farm advisor will need to assess the present effluent management, types of effluent, present storage facilities, the economics of developing a FCW, and the likelihood that a FCW will improve existing conditions. When assessing the type and volume of effluent to be treated, the present and possible future loading (*e.g.*, increased stock numbers, impervious areas, and shed roofs) should be considered.

Precipitation events are the primary factor affecting design, as there is almost no attenuation of farmyard-intercepted rain. The hydraulic flux in any year arising from storm events can be between 25 and 100 times the mean flow from the farmyard. It is this flux that must be appropriately managed within the FCW design.

Most FCW are designed to deal with storm events and the variable composition of influents through including sufficient wetland area for flow attenuation and water treatment. The areas of yards, tracks, and roofs within the farmyard must be calculated to determine the volumes to be treated. Each farmyard will have varying daily water usages (*e.g.*, yard washings and parlor washings), which must be calculated as part of the total volumes of water and runoff produced on site (Carty *et al.* 2008).

Before implementing a FCW, it is necessary to ensure that BMP are employed to decrease the contamination of the farmyard runoff (*e.g.*, improved cattle diet, roofing of feeding areas, and separation of roof runoff). It is also important to look at potential alternatives available to the farmer to deal with the contaminated farmyard runoff (*e.g.*, storage and subsequent land spreading (Bowmer and Laut 1992), and overland flow) as discussed elsewhere (Harrington *et al.* 2005; Scholz *et al.* 2007a, b).

Carty *et al.* (2008) provide a decision support tree for the treatment of farmyard and roof runoff with FCW. This tree can be used as a template and might require modification to address differences in regional and national legislation.

2.2.3.2 Site Characteristics

Each FCW design is site specific requiring "expert judgment" as discussed by Scholz (2006a, 2007). Therefore, a comprehensive site-specific assessment combining site investigation and desk study is necessary to determine the site characteristics, assess any potential impacts, decide on any groundwater and surface water protection measures, and provide data that will be used in the design of the FCW.

The site assessment should be undertaken by a person who is professionally qualified, has experience in all of the required disciplines, and can call in experts if necessary to clarify any anomalies that may arise. When assessing the site suitability, the relevant bodies and authorities should be consulted in the context of the prevailing legislation. The following existing characteristics of a site should be assessed: topography, geology, soils and subsoils, hydrogeology, hydrology, flora and fauna, archaeological and architectural features, and natural interest.

Climatic conditions, including rainfall and evapotranspiration, will determine the volumes to be treated by the FCW system and the residence time within the FCW. Due to the sources of effluent (mainly precipitation-generated runoff) and the effects of climate, discharge from the FCW may be confined to wet periods.

Ideally, an FCW should be developed on gently sloping land (Harrington *et al.* 2005). Areas that are steeply sloping require larger wetland areas, deeper soils and subsoils, more excavation, and probably increased costs of construction. When assessing the proposed site, consideration should be given to the approximate area of land required for the wetland, embankments, and access (see below).

The location of a wetland down gradient from the farmyard will allow for the effluent to flow by gravity and thus remove the need for pumping. Existing Irish FCW have been located at varying distances from 5 m to greater than 500 m from the farmyard. The FCW cells and embankments should be designed so that water flows through the FCW by gravity (Scholz *et al.* 2007).

A topographical survey of the farmyard and the proposed FCW site, at a scale of at least 1:500, is recommended. The survey should include contours (0.5-m contours), location and use of buildings, boundaries, hydrological features, and any features of archaeological and architectural interest.

A desk study should be undertaken in advance of the site visit to establish the geological context of the site. A number of trial holes should be dug on the proposed site to examine the type, depth, and texture of the underlying soil and subsoil. Trial holes should be dug to a depth of between 2 and 3 m, with a minimum of four trial holes per approx. 4000 m^2. Where the soil texture is not easily determined by hand assessment methods on site, a laboratory test can be conducted to determine soil permeability, particle size distribution, or both.

Where low-permeability soils suitable for the construction of a wetland (such as clays) are not found on site, alternative measures should be taken (Carty *et al.* 2008):

- Import suitable soils: there may be a requirement to import soils for part or all of the FCW. Additional soils can sometimes be found on the farm or locally, but this may incur additional costs for transportation, particularly if located away from the farm. Importing soil from another site or farm may require a license subject to national guidelines and laws.
- Use of an artificial liner: the use of an artificial plastic liner, similar to that required for a landfill, can also be considered. However, its use will incur a much greater cost and may require replacement in the future. Furthermore, the installation of an artificial liner requires the expertise of an engineer and consultation with the local authorities and regulators.

Field drains are widely used in agricultural land. All field drains must be located (through local knowledge or through site excavation) to assess whether they are likely to conduct water to the FCW or provide a pathway for water to leave the

FCW before completing treatment. A trench or drain dug during construction around the perimeter of the proposed FCW site to a depth of at least 1 m should intercept any field drains and divert their flow to surface water.

The types of flora, fauna, and habitats on and adjacent to the site should be described to determine whether the development of an FCW will have any negative impacts (*e.g.*, removal of hedgerows, nesting areas, and trees). The positive contribution of the FCW in creating new habitats or enhancing existing habitats and biodiversity should also be considered (Froneman *et al.* 2001).

The proximity to dwelling houses must be determined to assess issues such as security, health, safety, and odors. In some instances, access to the FCW by humans or livestock may need to be restricted by appropriate fencing or hedging, although this may only be required around the initial FCW pond.

The location and type of any archaeological or architectural feature on or adjacent to the site must be assessed, subject to national guidelines, using a combination of site investigation and desk study (Scholz 2007; Carty *et al.* 2008).

Carty *et al.* (2008) show a decision support tree for finding a suitable site for an FCW. This tree may require alteration depending on regional and national differences in managing agricultural land.

2.2.3.3 Discharge Options

Any nearby surface water features such as rivers, streams, lakes, and drains should be noted during the survey. In most cases, a local river, stream, ditch, or even woodland will provide the final discharge point for the water that has been treated by the FCW. As well as assessing the suitability of a watercourse as a potential discharge point, consideration must be given to high water levels, other discharges upstream and downstream, water abstractions downstream, and flooding. Where an FCW is to be developed in a floodplain, the potential impacts upstream and downstream must be assessed. In some instances, the floodplain may be protected, in which case an FCW may not be permitted. Regional and national authorities need to be consulted (*e.g.*, UNEP 2003; Zedler 2003; Scholz 2006a).

The final discharge should be to surface water with sufficient assimilative capacity, such as a stream or river with significant flows throughout the year, rather than to a field drain that has low assimilative capacity and often dries out. The discharge of the treated waters should have a negligible effect on the receiving water (*e.g.*, Zedler 2003).

Where a discharge to surface water is not possible or not suitable, a wet woodland or willow bed could be used. The woodland can be designed to have a zero discharge or minimal discharge; however, it must be noted that this may require a large area (Scholz 2006; Zheng *et al.* 2006).

It is recommended that an FCW should not be immediately adjacent to surface water to minimize the impact of any failure of the FCW. The buffer distance depends on the adjacent surface water and the ground conditions; a minimum of 5 to 10 m is recommended.

During the site assessment, information on the following should be collected: wells, springs, water table elevation, aquifers, nearby surface and groundwater supplies, and connectivity with surface water features, since they will cause increased volumetric loading to the system, reducing residence time and treatment. Furthermore, sufficient distance must be provided between an FCW and up-gradient and down-gradient water supplies (Scholz 2006a; Zedler 2003).

The height of the water table must be recorded. The base of the FCW cells should be above the water table, so sites with high water tables will not be suitable. The vulnerability of groundwater to pollution is assessed through combining information from groundwater vulnerability maps (where available) with a site-specific study and investigation of soil and geological conditions.

2.2.4 Design Guidelines for Farm Constructed Wetlands

2.2.4.1 Background and Water Treatment Requirements

Once site assessment and selection has been completed, the detailed design of an FCW can be conducted, taking into account the farm and farmyard structure and management practices. When designing a FCW, the following aspects are therefore of major importance (Zedler 2003; Harrington *et al.* 2005):

- objectives behind the construction of the FCW;
- characteristics of the farmyard runoff to be treated (volume and quality);
- water quality targets to be achieved; and
- land availability to achieve the target water quality.

Adequate pretreatment, retention time, management, and operation (*e.g.*, removal of sediment and regular inspection) and design for management (*e.g.*, access) are required to achieve effective water treatment through a FCW. A FCW that is fit for a certain purpose should have the following attributes (Harrington and Ryder 2002; Rice *et al.* 2002; Scholz *et al.* 2007a, b):

- be reliable and efficient in water treatment, particularly during storm events, extreme rainfall with increased hydraulic loadings, and also under relatively cold conditions;
- be capable of coping with accidental spillages;
- be flexible and versatile;
- be relatively simple to build;
- have low operation and maintenance requirements and costs;
- have low energy consumption;
- be a good landscape fit;
- enhance habitat and biodiversity (Froneman *et al.* 2001); and
- be safe for farmers and for the public.

2.2.4.2 Runoff Capture and Conveyance

One of the early steps in FCW construction should be to ensure that any contaminated runoff, such as from roofs, farmyards, and tracks, be captured properly. Runoff from adjacent land should usually not enter the FCW. However, regional recommendations may vary.

The conveyance of waters to, within, and from the FCW must consider the following (Scholz 2006a): collecting water from the farm, conveying that water to the wetland, moving water within the wetland, and moving water out of the wetland. Where possible, the effluent should flow by gravity to minimize maintenance and energy costs.

It is essential that any containment be secure and that only water with acceptable concentrations of contaminants be discharged to watercourses or groundwater. The FCW embankments retaining the water flowing through the system must be sufficiently high to allow for the accumulation of sediment and detritus. The soil lining the base must adequately impede infiltration to protect groundwater (Dunne *et al.* 2005a, b; Keohane *et al.* 2005; Scholz *et al.* 2007b).

2.2.4.3 Hydraulics, Water Balance, and Residence Time

The periodic nature of precipitation and the interception and uptake of water by emergent vegetation, evaporation, and ground infiltration has the capacity to arrest water flow between the individual segments of a FCW. This creates a freeboard between the outlet level and the level of the water contained within an individual wetland cell. It also provides each wetland cell with 'additional' receiving hydraulic capacity before flow to the next segment can resume, thus enhancing the hydraulic residence time (Dunne *et al.* 2005a, b; Scholz *et al.* 2007).

The treatment effectiveness of surface-flow wetland systems in comparison to sub-surface-flow systems (Mantovi *et al.* 2003) is typically based on having appropriate hydraulic residence times, which depend very much on the specific site conditions (Harrington *et al.* 2005). In shallow, emergent, or vegetated wetlands, such as FCW, this depends on having sufficient functional wetland area with an appropriate length-to-width ratio and a high emergent vegetation density. The hydraulic effectiveness of the FCW can be maximized by the following measures (Scholz *et al.* 2007):

- segmentation of the wetland into a number of wetland cells of appropriate configuration (see below);
- avoidance of preferential flow;
- dense vegetation stand; and
- managing the water depth to ensure optimal functioning (Scholz 2007).

The velocity of the water flow through the FCW is determined by the volumetric flow and the cross-sectional area of the water channel. Minimizing the velocity enhances the settling of suspended solids and promotes a longer contact time with emergent vegetation whose surfaces support biofilms (Scholz *et al.* 2002; Kanta-wanichkul and Somprasert 2005; Scholz *et al.* 2007).

Wind and temperature gradients can generate water movement between the different aquatic strata within a wetland cell. Emergent vegetation minimizes mixing, thereby allowing the cleaner water to flow preferentially along the surface, especially during periods of large precipitation-generated flow. In the initial receiving wetland cell, floating vegetation may develop (typically *Glyceria fluitans* and *Agrostis stolonifera*) and water flow will be partially sub-surface, thus having the additional advantage of reducing odors (Dunne *et al.* 2005a; Scholz and Lee 2005; Scholz *et al.* 2007).

2.2.4.4 Wetland Sizing, Inlet, and Outlet

The design and sizing of FCW has often focused on phosphorus, which is recognized as one of the most difficult contaminants to remove from water and is a limiting nutrient in many freshwater ecosystems (Braskerud 2002). For example, a catchment-specific study of 13 wetland systems in Waterford (Ireland) showed that, to achieve a mean MRP concentration at the outlet of 1 mg/l or less, the wetland area required was at least 1.3 times the farmyard area, and that each system should contain approximately four cells.

The design is based on two assumptions: the larger a wetland, the more phosphorus removal can be expected; and all ICW studied near Waterford (Ireland) were at the designated threshold of failure for phosphorus (1 mg/l MRP for the outflow or near it in cases of no flow (Scholz *et al.* 2007). This finding relating to MRP is, however, not universally applicable.

The aspect ratio is defined as the mean length of the wetland system divided by the mean width. The study conducted in Ireland showed that to obtain an outlet MRP concentration of 1 mg/l or less, the FCW aspect ratio should be less than 2.2. In fact, the closer the aspect ratio is to 1 (*i.e.*, the more the FCW shape is square or round), the better the wetland treatment (Scholz *et al.* 2007).

Sizing must also take into account the footprint of the upper embankments, which should be between 2 and 3 m wide to ensure stability and to provide easy access for maintenance and monitoring. For safety reasons, inner embankments should be gently sloping.

Inlets and outlets should be kept as simple as possible and avoid the use of concrete and overengineered structures. Pipe diameters should be at least 150 mm to avoid clogging. Stone chippings should be placed beneath the inlet and outlet pipes to prevent scouring. Elbow pipes fitted to linear ones can be used to control the water level and the outflow of each cell. Carty *et al.* (2008) provide further details on sustainable wetland design.

2.2.4.5 Landscape Fit, Biodiversity, and Life Span

The potential visual aspect of the FCW system design is important for achieving empathy from both farm dwellers and the local community. Usually, FCW with curvilinear shaping and virtually level embankments have a more 'natural' appearance.

Several measures can improve the landscape fit and biodiversity of an FCW. Through BMP on the farm, the level of contaminated water discharging to the ponds can be reduced (Scholz 2006a; Zheng *et al.* 2006). Wherever possible, the FCW should be located near (but not connected to) existing wetlands, ponds, and lakes to allow for natural colonization by plants and animals. The FCW cells should be irregular in shape, with gently sloping embankments and areas of deeper water, and contain islands where sufficient area is available. The use of locally occurring wetland plant species for establishing habitats and enhancing biodiversity appropriate to the locality is also likely to further increase the robustness and sustainability of the system (Froneman *et al.* 2001; Scholz 2007; Scholz *et al.* 2007b).

The area surrounding the FCW can be planted with trees and shrubs, but trees are not recommended on the FCW embankments. If possible, small pools around the main system should be created to collect runoff from adjacent fields and create additional aquatic habitat (Froneman *et al.* 2001; Carty *et al.* 2008).

Wetland embankment height, inflowing solids, and accumulating detritus determine the functional life span of each segment of the FCW. With detritus accumulation and a minimum embankment height of 1 m, a life span of between 50 and 100 years is expected. However, the life span can be virtually indefinite if detritus removal takes place regularly, as discussed by Scholz *et al.* (2007b) and in the section on maintenance below.

2.2.5 Construction and Planting

2.2.5.1 Construction

Ideally, the construction of an FCW should be undertaken during the dry season and the involvement of the farmer is encouraged. Construction of an FCW should be undertaken by a competent machine operator and signed off by a qualified engineer on completion, although regional and national requirements for construction vary. The main stages of construction are as follows (McCuskey *et al.* 1994; Scholz *et al.* 2007a, b):

1. Stripping of topsoil from FCW area and retention of it for later use (if applicable);
2. excavation of subsoil;
3. layering and compaction of soils for cell liner (laying of an artificial liner if required);
4. creation of gently sloping embankments (at least 1 in 3), with height >1.0 m and tops 2 to 3 m wide to guarantee stability and sufficient access around the wetland;

5. potential redistribution of topsoil (if nutrient poor) over the base of the ponds;
6. pipe laying between cells;
7. placement of stones or chippings beneath inlet and outlet pipes; and
8. planting of FCW cells.

The machinery required for construction comprises a tracked excavator and bulldozers and a vibrating roller where a higher degree of soil compaction is required. The topsoil is stripped from the FCW area and retained for later use. The depth of excavation will vary depending on the depth of the overlying topsoil (McCuskey *et al.* 1994; Carty *et al.* 2008).

Subsoil is excavated to form the base of each pond and to provide material for the embankments. The depth of excavation will depend on the topography and elevations required between each pond; typically, an excavation of 0.5 m below the ground level is required for flat sites. Where surrounding land slopes towards the FCW system, a drain or an embankment should be constructed around the system to divert potentially large volumes of runoff away from the system (McCuskey *et al.* 1994; Carty *et al.* 2008).

The amount of compaction and layering will depend on the type of material being used to construct the FCW and should be determined during the site assessment of the soils. Where the soil permeability is sufficiently low, no layering is required. Medium- to low-permeability soils will require layering and compaction to ensure a permeability of approx. 10^{-8} m/s. Regional and national guidelines on soil permeability thresholds should be consulted by the designers.

Pipe ducting and elbows are placed at the inlet and outlet points for each FCW cell (see above) to ensure that the movement of water in each cell is across the maximum distance from the point of inflow to the exit. Furthermore, piping should be positioned in accessible locations and placed as low as possible to the base of the exit point of the upper cell to ensure the possibility of drainage. In FCW discharging to surface waters prone to flooding, a non-return valve should be placed on the outlet pipe to prevent any flood water from entering the wetland system. Stones or chippings should be placed beneath the inlet and outlet pipes to prevent scouring of the wetland floor and to provide access for monitoring (McCuskey *et al.* 1994; Carty *et al.* 2008).

The banks and floor of the wetland cells should normally be compacted and smoothed off using tracked excavators that are suitable for use on difficult wet terrain. Where a greater degree of compaction is required, rollers are used.

Measures should be employed during the construction of an FCW to limit the impact on surface water (runoff and siltation) and groundwater through proper management and supervision following national guidelines.

Depending on the location of the site, there may be a need for fencing to restrict access to humans and livestock. When determining the type of fencing, the farmer will need to consider the extent of fencing, costs, and human and machinery access. In most cases, fencing similar to that already used on the farm will be suitable (Carty *et al.* 2008).

2.2.5.2 Planting

The primary vegetation used in FCW is composed of emergent aquatic plant species (helophytes). These species have evolved to enable them to root in soils with
little or no available oxygen, growing vertically through the water column with
most of their leaves in the air. They have specially adapted tissues and physiolo-
gies that facilitate oxygen storage and its transportation from the leaves through
the stem to the roots. The soil and water characteristics influence the type of helophyte species and their treatment performance (Scholz 2006a; Scholz *et al.* 2007).
Helophytes in FCW have important functions such as the following (Scholz and
Lee 2005; Jiang *et al.* 2007):

- provision of a support structure for microbial colonies to develop;
- facilitation of aerobic microbial activity (principal treatment removal process);
- uptake of nutrients;
- source of organic matter;
- reduction of water flow to increase residence time and settlement; and
- reduction of final volumetric discharge through plant transpiration and interception.

While more than 100 native helophyte species can be used, about 20 species are
actively planted in constructed wetland systems (Scholz and Lee 2005; Scholz
2006a). Initial plant establishment is the dominant influence on the vegetation
structure on an FCW during its early years. Plant establishment is dependent upon
the nature of the influent and water depth, plant species and physiological maturity, and planting density (Scholz *et al.* 2007b). Interspecies competition and other
biotic factors, especially waterfowl, influence long-term vegetation development
(Froneman *et al.* 2001; Scholz 2006a).

Plants should be sourced from existing (constructed or natural) wetlands where
permitted or from accredited plant nurseries. Care must be taken to ensure that
non-native species are not introduced. Seedlings are usually not recommended as
they are more vulnerable to pollution and water level variations and take longer to
establish (Carty *et al.* 2008).

If plants are sourced from existing wetlands, sufficient root material should be
obtained to allow the plants to establish, while also ensuring that the root system
of the original plant can regenerate. Harvesting should be carried out ideally by
hand and over a sufficiently large area to minimize disturbance to the plants and
its environment.

Planting of mature macrophytes should be carried out by hand into water or
suitably saturated soils, ideally with 50 to 100 mm of water above the topsoil.
Water levels should be maintained between 100 and 200 mm for at least the first 6
weeks after planting, and the wetland should not be allowed to dry out below the
soil surface. Shallow waters (<100 mm) will encourage establishment of grasses
and weeds, which can restrict the growth of the wetland vegetation. Water may
need to be sourced from nearby surface water features during planting, as the

effluent from the farmyard may only provide sufficient cover for the first cell (McCuskey *et al.* 1994; Scholz and Lee 2005).

The area of each wetland should be planted with various wetland species to increase biodiversity. The first wetland cell will receive a higher pollution load, often with a high ammonia–nitrogen concentration, so a minimum of three plants per square meter is recommended to increase plant success. Subsequent cells will have reduced effluent loadings, and a minimum of two plants per square meter is therefore recommended. The farmer may wish to allow areas within the final cells to colonize naturally. Provided that planting is carried out at the beginning of the growing season and water levels are maintained between 100 and 200 mm for the initial few months, a desirable vegetation cover of at least 80% after 2 years should be attained. To establish a dense cover in a shorter period, the initial planting densities should be doubled (Carty *et al.* 2008).

Prior to the commissioning of the FCW, the system should be signed off by a competent engineer (depending on regional guidelines) and have all monitoring and maintenance features and fencing in place. During the first few months, water levels within the wetland ponds should be at a minimum, approx. 100 to 200 mm, to provide favorable conditions for plant establishment. It is possible that some effluent may need to be diverted away from the FCW during the start-up stage; *e.g.*, stored or land spread (McCuskey *et al.* 1994; Carty *et al.* 2008).

2.2.6 Maintenance and Operation

2.2.6.1 Pipe Maintenance and Flow Control

The success of a FCW will depend on the maintenance and operation of the system. While a FCW is designed to be as self-maintaining as possible, it is crucial that a maintenance program be adopted to ensure continued effective water treatment and 'rejuvenation of the system'.

All inlet and outlet pipes within the FCW system should be visually inspected weekly for blockages, sediment accumulation, and debris. Blockages will affect the hydraulics of the FCW system, while sediment accumulation may indicate inadequate solids separation further up in the system. Any blockages and sediment or debris accumulations around the inlet or outlet pipes should be cleared (Harrington *et al.* 2005; Scholz 2006a).

During prolonged dry periods, water depths within the ponds will decrease, especially in down-gradient regions of the wetland. It is essential to ensure that soils are flooded (to at least 50 mm). Usually, FCW should not be allowed to dry out as cracks may form in the base, which may cause higher infiltration rates in the short term when effluent reenters the cell. Once the emergent vegetation is established, FCW in temperate regions should be able to cope with reduced water depths, and even drying out, for periods of several months.

Any adjustment of pipes must be carried out gradually as these movements may cause huge surges of effluent to subsequent wetland cells or receiving surface or groundwaters, reducing residence time and treatment within the system.

2.2.6.2 Vegetation and Sediment Maintenance

Water levels should be maintained at less than 300 mm to ensure good plant growth. However, most macrophytes tolerate short periods of increased water depth (up to 500 mm), such as that associated with high rainfall, as well as low water depths or even no water during dry weather. Major changes in vegetation cover and composition should be noted as a possible indicator of changes (*i.e.*, degradation or improvement) in wetland performance. Pest control might be required if unwanted plant species take over a wetland.

Sediment accumulating within a wetland comprises organic material from the influent and dead plant matter, and also mineral sediment from eroded tracks. Existing constructed wetlands have shown varying accumulation rates depending on the type of loading and the amount of vegetation cover within the wetland. The removal of accumulated sediment is usually confined to the first cell (Scholz *et al.* 2007).

For a heavily loaded system, the inclusion of an open water pond at the initial stage of the FCW as a sediment trap may extend the operational life of subsequent cells before the removal of material is required. Any pond located within the initial wetland cell would require relatively frequent material removal (most likely biannually) but could be configured to be as accessible as a standard slurry storage tank (Scholz 2006a; Scholz *et al.* 2007b).

The most appropriate way of managing the material removed is likely to be land spreading on the farm in accordance with good farm management practice. Information on the solid content and nutrient composition, particularly phosphorus, is required to ensure that the usage complies with farm nutrient management requirements (Bowmer and Laut 1992; Carty *et al.* 2008).

As the farm's nutrient management requirements may not allow the full amount of removed sediment to be spread on the land for one year, it is likely that occasional storage of removed sediment would be required for some sites. The storage requirements specified in good agricultural practice regulations with respect to farmyard manure should also suffice for FCW sediment to ensure environmental protection.

2.2.6.3 Safety, Security, and Maintenance

It is recommended that the farmer undertake regular visual inspection of the internal and external faces of the wetland embankments to check for any water leakage, slippage, or distortion. The internal embankment face can be checked by walking along the embankment crests and external embankment faces by walking along the external boundary of each cell. Any defects such as leakages, slippages, or distorted areas should be addressed immediately.

Access around the wetland should be maintained by managing vegetation growth on the embankments. Under normal operating conditions, growth on some pond crests will need to be cut biannually using a mower or topper. Security and safety considerations for both humans and livestock should be incorporated into the design of the FCW subject to national guidelines. Access to the influent-receiving segment of the FCW may have to be limited through fencing or by existing hedgerows around the FCW.

2.2.6.4   Monitoring the Final Effluent and Receiving Watercourses

The monitoring of the FCW and the final effluent will allow the farmer to assess the performance of the wetland system and to detect any malfunction. The general appearance of the final effluent should be noted, paying particular attention to water color, the presence of "sewage fungi" (Scholz 2006a), smell, and any evidence of plant material in the discharge. If the final discharge water appears to be heavily discolored or polluted or contains plant material, then the outlet pipe should be isolated immediately by closing the gate valve. However, water that is visibly clear may also have a high nutrient load, which can only be determined by laboratory analysis.

The condition and appearance of the receiving waters at the point of discharge should be checked on a monthly basis and following extreme events such as high rainfall. The farmer should assess the condition and appearance of water, both upstream and downstream of the discharge location. Heavily discolored water or the appearance of sludge-type material may indicate an upstream pollution source. Foaming immediately downstream of the discharge point may indicate pollution in the final effluent. The outlet pipe from the FCW should be isolated immediately by closing the gate valve in the event of any suspected pollution incident. The farmer should obtain advice from a suitable agricultural advisor and the regulator (Scholz 2006a; Carty et al. 2008).

2.2.7   Conclusions

This sub-section proposes, for temperate climates, universal design, operation, and maintenance guidelines for a FCW, which are a specific application of ICW. Guidelines have been proposed for assessing the need for FCW and their site suitability. The overall design of FCW is empirical, site specific, and predominantly based on expert judgment. The guideline is based on minimizing costs for the benefit of the farm owner but not at the expense of the environment. Therefore, FCW may require considerably more land than conventional treatment technologies, which are usually associated with greater maintenance and capital costs.

2.3 Integrated Constructed Wetland for Treating Farmyard Runoff

2.3.1 Introduction

Farmyard runoff, which is rich in nutrients, is a source of diffuse pollution and potentially a serious risk to receiving watercourses by contributing to eutrophication (Cleneghan 2003). Constructed wetlands have been used worldwide to treat different categories of wastewater including domestic, industrial, acid mine drainage, agricultural runoff, and landfill leachate (Kadlec and Knight 1996; Scholz 2006a).

The ICW concept (Harrington *et al.* 2007; Scholz *et al.* 2007a, b) is based on an approach that endeavors to achieve water treatment, landscape fit, and biodiversity enhancement targets by an innovative wetland design methodology. One subgroup of ICW is the FCW as defined by Carty *et al.* (2008).

The conventional practice in Ireland is land spreading of farmyard dirty water (Tunney *et al.* 1997). The storage and spreading is governed by rules to prevent water pollution. However, this practice requires considerable labor and machinery resources as well as storage infrastructure. Improper storage and spreading has been linked to water pollution, particularly to increased levels of nitrogen and phosphorus in surface and ground waters (Healy *et al.* 2007). In contrast, the ICW concept is founded on the holistic use of land to protect and improve water quality. These systems are areas of land–water interface that form an integral part of the environmental and ecological structure of the landscape. They act as buffer lands that control the transfer and storage of farmyard dirty water rich in nutrients (Scholz *et al.* 2007b).

The main characteristics of an ICW, such as shallow water depth, emergent vegetation, and the use of *in situ* soils, mimic those found in natural wetland ecosystems (Harrington *et al.* 2005). Scholz *et al.* (2007a, b) reported on the detailed concept of these synergistic, robust, and sustainable systems by referring to case studies in Ireland.

Contaminated effluent is treated in an ICW through various physical, chemical, and biological processes involving plants, microorganisms, water, soil, and sunlight (Kadlec and Knight 1996; Scholz 2006a). The extent of treatment in constructed wetlands depends upon the wetland design, microbial community, and types of aquatic plants involved. Water quality improvements are predominantly caused by bacteria (Ibekwe *et al.* 2003); for instance, most of the biological degradation takes place within bacterial films present on sediments, soils, live submerged plants, and the associated litter. Microbes catalyze chemical changes in wetland soils and indirectly control the nutrient availability to plants and, in turn, the water quality (Mitsch and Gosselink 2007). The litter and associated sediment originating from decaying macrophytes provide considerable surface area for the attachment of biofilms and are therefore important for microbial processes such as the transformation of nutrients in wetlands (Scholz 2006a).

Some of the main processes for nitrogen transformation in constructed wetlands are nitrification, denitrification, and anammox (Shipin *et al.* 2005; Vymazal 2007), which are all microbe-mediated processes (Wallace and Knight 2006). It is expected that nitrogen removal will fluctuate and increase over time as vegetation becomes established and sufficient carbon is available for denitrification (Kadlec and Knight 1996).

Wetlands have a relatively high capacity to store nutrients (Braskerud 2002). Phosphorus assimilation in constructed wetlands depends on factors such as the influent phosphorus concentration, the rate of internal biomass cycling, and the wetland age (Kadlec 1999; Wallace and Knight 2006). However, the long-term storage of phosphorus in these systems is linked to the cycling of phosphorus through the growth, death, and decay of plant biomass. Previous studies on wetland ecosystem structure and function have shown that soil is the most important long-term ecosystem phosphorus storage compartment (Richardson and Marshall 1986). Processes such as soil adsorption and peat accumulation control the long-term sequestration (*i.e.*, capture) of phosphorus (Richardson and Marshall 1986; Reddy *et al.* 1999).

Most previous studies have been short term, and on either pilot plant- or laboratory-scale experimental systems. Very few studies have been conducted on full-scale constructed wetlands with long-term evaluations (Brix *et al.* 2007; Newman *et al.* 1999). There is a lack of information on the performance of mature constructed wetland and ICW systems including FCW structures. The purpose of this study was therefore:

- to evaluate the treatment efficiency of a full-scale mature ICW system that has been in operation for 7 years;
- to assess the annual and seasonal variations in nutrient removal;
- to identify the presence of microorganisms responsible for nitrogen transformations within these systems; and
- to investigate the impacts of potential contamination of nearby surface waters and groundwater.

2.3.2 Material and Methods

2.3.2.1 Site Description

The researched ICW treatment system is a FCW, situated in County Waterford (southeast of Ireland) at a longitude of 07°02′40″ W and a latitude of 52°11′28″ N (Figure 2.14). The case study area is located in a temperate zone with a mean annual temperature and precipitation of 11.4°C and 1094 mm, respectively (Met Éireann 2007). Mean seasonal temperatures for the region in 2007 were as follows: winter, 7.8°C; spring, 10.3°C; summer, 14.9°C; and fall, 12.2°C.

The ICW system was constructed in 2000 and commissioned in February 2001 to contribute to the improvement of the water quality of the Annestown Stream.

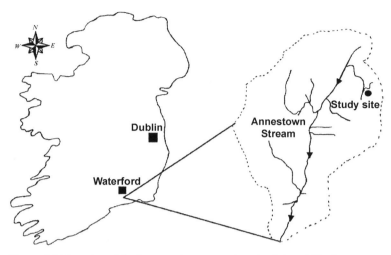

Figure 2.14 Location of the integrated constructed wetland site within the Annestown catchment near Waterford, Ireland

The ICW system has a total area of 0.76 ha. The primary vegetation types used in the ICW are emergent plant species (helophytes). The first three cells were densely vegetated, while the last cell had only sparse vegetation.

The cells were not lined with an artificial liner. However, the subsoil was reworked and used as a natural liner. The cells were only partly planted with vegetation naturally available on the site. Further plant growth occurred by natural recruitment. Each cell had one inflow and one outflow structure. The water flow between each cell was by gravity through a PVC pipe.

The effluent entering the ICW system comes from a dairy farm of 0.5 ha with 77 cows. The wastewater contained farmyard and roof runoff occasionally contaminated by manure and was conveyed to the ICW system by gravity through a pipe.

The key features of this constructed wetland were horizontal surface flow and intermittent hydraulic loading. The ICW system was based on four cells operated in series. A single influent entry point was located at the first cell. The first three cells had depths of approx. 1 m, while the fourth cell was approx. 1.5 m deep.

2.3.2.2 ICW Design

Information on topography, hydrogeology, surface water, groundwater, and subsoil is important for assessing the ICW site suitability prior to the design and construction phase. Details concerning site assessment criteria have been documented by Scholz *et al.* (2007b).

The ICW system was designed not only to improve water quality but also to integrate the structure naturally into the landscape and enhance biodiversity. The design philosophy is in agreement with the guidelines proposed by Carty *et al.* (2008).

A detailed site-specific assessment combining site investigation and desk study was conducted to determine the site characteristics and assess any potential impacts on groundwater and surface waters. The site was assessed by characteristics such as topography, geology, soils and subsoils, hydrogeology, hydrology, flora and fauna, archaeological and architectural features, and natural interest.

A topographical survey of the farmyard and the proposed ICW site was conducted that included contours (0.5 m contours), location and use of buildings, boundaries, and hydrological features. The nearby surface water features such as rivers, streams, and drains were also noted during the survey. The suitability of the nearby stream as a potential discharge point was also assessed.

During the site assessment, information on wells, springs, water table elevation, aquifers, nearby surface and groundwater supplies, and connectivity with surface water features was also gathered. The soil and geological conditions were also studied.

2.3.2.3 Sampling and Analytical Methods

Grab samples for each wetland cell inlet and outlet were taken on an approximately fortnightly basis. Liquid samples were analyzed for variables including the 5 d at 20°C N-allythiourea biochemical oxygen demand (BOD), chemical oxygen demand (COD), suspended solids (SSs), ammonia–nitrogen, nitrate–nitrogen, molybdate reactive phosphorus (MRP; equivalent to SRP), total coliforms (TCs), and *E. coli* at the Waterford County Council water laboratory using predominantly American Public Health Association standard methods (APHA, 1998) unless stated otherwise.

Continuous flow measurements were undertaken between April 2003 and June 2004. From June 2004, spot measurements were made of flows into and out of each ICW cell. Furthermore, flows entering and leaving ICW cells were monitored and recorded using standard Greyline flow meters and associated data loggers. Blockages due to vegetation required frequent maintenance of the flow meters.

Four piezometric groundwater-monitoring wells were placed up-gradient (one well), within (two wells), and down-gradient (one well) at different depths at the ICW site in fall 2004. These wells were sampled on a quarterly basis throughout 2004 and 2007. The day before sample extraction, all wells were purged (*i.e.*, process of removing water that is unrepresentative of the surrounding strata) prior to sampling, and subsequent analysis was carried out according to APHA (1998).

The water quality of the receiving stream, which is a tributary of the Annestown Stream, was monitored (Figure 2.14). Grab samples were collected from 21 locations along the stream located adjacent to the ICW. Three points were sampled upstream, up to 16 parallel to cell 4, and 2 points downstream to assess the effect of the ICW system on the associated receiving watercourse. In 2007, grab samples were taken predominantly during an intense period for monitoring the water quality of the receiving watercourse. The sampling scheme was designed to monitor the water quality during periods of relatively low and normal flows.

Weekly samples were collected in late spring and during the following summer. Later on, monthly samples were collected during fall and winter. Nutrient analysis was conducted at the Waterford County Council water laboratory using American standard methods (APHA 1998).

In May 2007, sediment samples were collected with a sediment sampler (diameter of 4 cm) from the wetland cells to identify bacterial groups responsible for nitrogen transformations in the ICW. Predominantly triplicate field sediment samples were collected 1 m to the left, 1 m in front, and 1 m to the right of each sampling point located near the inlet of each ICW cell and outlet of the last ICW cell to reduce bias and to gain more representative samples. Samples were stored at −20°C before analysis.

All samples were frozen immediately after collection and sent off to Linköping University (Sweden) for subsequent molecular microbiological analysis involving standard techniques (*e.g.*, extraction of deoxyribonucleic acid (DNA), polymerase chain reaction (PCR), and gel electrophoresis) as outlined by Sundberg *et al.* (2007).

Sediment samples were subjected to DNA extraction using a FastDNA SPIN kit for Soil (Bio 101, La Jolla, CA). Samples (0.25 g) were suspended in a sodium phosphate buffer supplied with the kit as stipulated by the manufacturer and homogenized for 180 s with a handheld blender (DIAX 900 Homogeniser Tool G6, Heidolph, Kelheim, Germany).

Bead beating was undertaken to disrupt the soil aggregates and to lyse bacterial cells mechanically. Bead beating was extended to 3×30 s to achieve good homogenization of the samples. The subsequent centrifugation was extended to 2×5 min and the centrifugation after washing with SEWS-M, a salt and ethanol wash solution (Qbiogene, USA), was extended to 5 min. The extracted DNA was stored at −20°C.

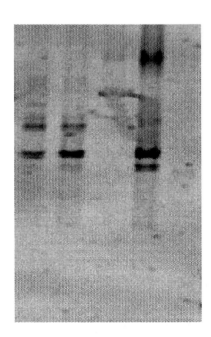

Figure 2.15 Denaturing gradient gel electrophoresis profiles of polymerase chain reaction products: ammonia-oxidizing bacteria for four example wetland cells

The ammonia-oxidizing bacterial community (Figure 2.15) was assessed using group-specific PCR primers, which were unique sequences from 16S ribosomal ribonucleic acid (rRNA) genes, while the denitrifying bacterial community (Figure 2.16) was assessed using primers that were unique sequences from the functional gene nitrous oxide reductase (nosZ). A primer is a nucleic acid strand that serves as a starting point for DNA replication and is required because most DNA polymerases (*i.e.*, enzymes that catalyze the replication of DNA) cannot synthesize *de novo* DNA. Ribosomal ribonucleic acid is one of the three major types of ribonucleic acid (RNA). Ribosomal ribonucleic acid and the genes that encode them are ideal biomarkers (Head *et al.* 1998).

The extracted DNA from all samples was diluted tenfold to avoid inhibition of the PCR by humic substances. The PCR amplification was undertaken using forward and reverse primers (CTO189f – GC A/B – GC, *CCGCCGCGCGGCGGGCGG-GGCGGGGGCACGGGGG*GAGRAAAGCAGGGGATCG; CTO189f – GC C, *CGCCCGCCGCGCGGCGGGCGGGGCGGGGGCACGGGGG*GAGGAAAGTA GGGGATCG; CTO654r, CTAGCYTTGTAGTTTCAAACGC) for ammonia-oxidizing bacteria as reported by Kowalchuk *et al.* (1997). The gas chromatograph clamps for primers are underlined (if used). The PCR was performed on a PTC-100 thermal cycler (MJ Research, San Francisco, CA, USA) in a 50-µL mixture.

The forward and reverse primers (nosZF, CG (C/T) TGT TC (A/C) TCG ACA GCC AG; nosZ1622R-GC, *GGCGGCGCGCCGCCC GCCCGCCCCCGTCG-CCC* CGC (G/A) A (C/G) GGC AA (G/C) AAG GT (G/C) CG) targeting the

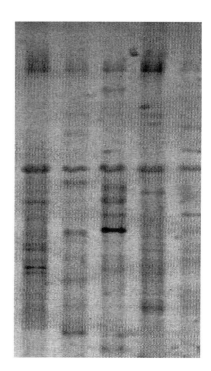

Figure 2.16 Denaturing gradient gel electrophoresis profiles of polymerase chain reaction products: denitrifying bacteria for four example wetland cells

nosZ gene (Throbäck *et al.* 2004) were used in a subsequent PCR. The GC clamps for all primers are underlined.

The PCR was again performed on a PTC-100 thermal cycler in a 50-µL mixture including 1.33 U of Taq polymerase and 5 µL of the supplied buffer (including 1.5 mM $MgCl_2$; Roche Diagnostic, Manheim, Germany), each nucleotide at a concentration of 200 µM, the primers at 0.125 µM each, 600 ng µ/l bovine serum albumin, and 2 µL of the DNA template.

The PCR products of DNA extraction and PCR reactions were examined by agarose gel electrophoresis. The agarose mixture (1% agarose in Tris acetate ethylenediaminetetraacetic acid (EDTA)) was melted by heating in a microwave for approx. 2 min and subsequently poured into an agarose gel casting tray. The solidified gel was covered with an electrophoresis Tris acetate EDTA buffer before running electrophoresis at 120 V for 40 min.

The PCR products and dye supplied with the DNA extraction kit (2µL of dye and 4µL of PCR products) were placed into loading wells. The first well of each row was loaded with 2µL of GeneRuler (1 kb DNA ladder; 1000 base pairs for ammonia-oxidizing bacteria and nitrous oxide reductase nosZ) and 4µL of distilled water. The electrophoresis was run for 40 min at 120 V (Owl Scientific, Woburn, MA, USA). The gel was then placed in ethidium bromide solution (immersed for 15 min), located in the fume cupboard, and subsequently washed with tap water. The ethidium-bromide-stained gel was then visualized by UV illumination.

2.3.2.4 Statistical Analyses and Limitations

All statistical analyses were performed using the analytical and graphical software tool Origin 7.5. The seasonal effects on water quality improvement were tested by ANOVA at $p < 0.05$.

There was no true experimental replication because all ICW are different from each other, and only one ICW was studied in detail. However, the experimental error was reduced due to an assessment of potential outliers and sub-sampling where required. The limitations associated with the statistical analysis of this case study suggest that the outcomes presented may only be valid for the specific ICW system assessed and very similar systems.

The ICW system studied is semi-natural and open. Considering that flow rates are therefore partly unknown and that flow estimations would result in very inaccurate constituent mass estimations, the display of constituent concentrations was chosen instead. This is common practice for less engineered and rather natural systems.

It has to be emphasized that ICW are not fully engineered treatment wetlands where the inflow and outflow rates are known and where losses to groundwater are zero. Moreover, ICW are purely driven by hydrology (*i.e.*, storm events) and not by rather constant inflow rates, which are common for constructed wetlands treating, for example, domestic wastewater.

2.3.3 Results and Discussion

2.3.3.1 Water Quality

The mean inflow values were as follows: BOD, 540.2 ± 697.13 mg/l; COD, 1501.7 ± 2007.07 mg/l; SS, 261.4 ± 1004.74 mg/l; ammonia–nitrogen, 39.62 ± 41.762 mg/l; nitrate–nitrogen, 3.81 ± 3.375 mg/l; MRP, 11.55 ± 10.049 mg/l; total coliforms, $1.3 \times 10^6 \pm 1.5 \times 10^6$ colony-forming units (CFU)/100 ml; *E. coli*, $9.6 \times 10^5 \pm 10.0 \times 10^5$ CFU per 100 ml; pH, 7.5 ± 0.90. These values indicate the very high variability of the farmyard runoff quality entering the ICW system (Mustafa *et al.* 2009).

The mean effluent water quality was as follows: COD, 75.5 ± 92.68 mg/l; BOD, 12.9 ± 10.60 mg/l; SS, 16.3 ± 17.61 mg/l; ammonia–nitrogen, 0.37 ± 0.562 mg/l; nitrate–nitrogen, 0.99 ± 1.812 mg/l; MRP, 0.94 ± 0.628 mg/l; TC, $1.0 \times 10^4 \pm 1.3 \times 10^4$ CFU per 100 ml; *E. coli*, 30.1 ± 34.4 CFU/100 ml; pH, 7.6 ± 0.55. The findings indicate that the ICW system had a high capacity to remove pollutants due to the large size of the wetland cells and the relative high mean retention time. Moreover, microbiological contaminants were removed well, most likely due to predation and unfavorable growth conditions at low temperature (Scholz 2006a).

For example, the overall reduction of 97.6% in BOD is similar to the 97% reported for a comparable system in the state of Maryland, USA (Schaafsma *et al.* 2000), and the 99% calculated for an ecological treatment system comprising a series of reactors, clarifiers, and wetlands in the state of Ohio, USA (Lansing and Martin 2006). Furthermore, the removal efficiency is higher in comparison to other wetlands treating similar types of influents: 65 and 76%, Connecticut, USA (Knight *et al.* 2000 and Newman *et al.* 1999, respectively); 40 to 50%, New Zealand (Tanner *et al.* 1995); 80%, Oregon, USA (Skarda *et al.* 1994).

Concerning other parameters, the COD and SS removal efficiencies were also high: 94.9% and 93.7%, respectively. This can also be explained by the large size of the wetland cells and the high retention times. The mean pH value increased slightly between influent and effluent from 7.5 to 7.6.

2.3.3.2 Comparison of Annual Treatment Performances

There were no significant annual trends in mean ammonia–nitrogen and MRP reduction between 2001 and 2007. Concerning nitrate–nitrogen, however, the mean reduction decreased by 6.2% in 2007, which was, however, statistically not significant. The MRP reductions were similar to the SS reductions, which might indicate that phosphorus is often bound to SS within the inflow water (Scholz 2006a; Mustafa *et al.* 2009).

As the ICW system continued to mature, the microbial communities (see below) and aquatic vegetation became more established, resulting in a stable system with high pollutant removal capacity. The overall nutrient removal efficiency of the system was high: 99.6% for ammonia–nitrogen, 86.8% for nitrate–nitrogen, and 93.2% for MRP, even in the sixth year of its operation. This finding contrasts with the common notion that the nutrient removal efficiency of constructed treatment wetlands decreases with age, especially for phosphorus removal as the mineral sediment becomes fully saturated; *i.e.*, no free adsorption sites remain (Kadlec 1999; Kent 2000). This may imply that ICW cannot be compared with traditional treatment wetlands in terms of their capacity to retain nutrients.

The characteristics of soil have an important influence on the magnitude of phosphorus losses. Different factors such as particle size distribution, organic content, and iron and aluminum concentrations influence the ability of wetland soil to hold phosphorus (Sharpley 1995; Leinweber *et al.* 1999; Daly 2000). In general, soils with high clay content have a higher capacity to bind phosphorus than those with sandy and organic soils (Sharpley 1995; Maguire *et al.* 1997; Leinweber *et al.* 1999; Daly 2000).

Also soils with higher iron and aluminum content were found to have a greater capacity to bind phosphorus (Daly 2000). Dunne *et al.* (2005a) demonstrated with the help of an intact soil/water column study that the phosphorus sorption parameters were significantly related to the amorphous forms of iron and aluminum oxides in soils. The soils containing about twice as much amorphous forms of iron and aluminum had a higher phosphorus sorption capacity (1601 ± 32 mg/kg as compared to 674 ± 62 mg/kg). Previous research conducted by Dunne *et al.* (2005a, b) showed that the soil in the current study site area contains high proportions of sand, silt, and clay ($12 \pm 0.3\%$ sand, $54 \pm 2.0\%$ silt, and $33 \pm 1.7\%$ clay), and most likely also relatively high amounts of iron and aluminum, resulting in a greater phosphorus binding capacity of the soil.

Furthermore, as the wetland ages, the continuous accumulation of withered plant biomass increases the phosphorus storage capacity of the ICW system. Moreover, the wetland plants were not harvested, resulting in the accumulation of organic matter (Seo *et al.* 2005). The wetland subsequently changes from an initially mineral-based system to an organic-based system with higher phosphorus removal capacity.

2.3.3.3 Seasonal Performance

Seasonal variations in performance concerning nutrient removal were observed. The ammonia–nitrogen concentrations within the inflow were relatively high in summer and fall compared to winter and spring. In comparison, the ammonia–nitrogen outlet concentrations were higher during fall compared to other seasons. It follows that the mean removal efficiencies were higher during spring (99.4%) and summer (99.7%) and slightly lower during fall (98.7%) and winter (99.3%). Although the mean removal efficiencies in fall and winter were relatively high, the mean concentrations of the inflow drastically increased during these two seasons and the mean concentrations in the outflow were 0.92 ± 0.769 and 0.60 ± 0.201 mg/l, respectively. The ammonia–nitrogen concentrations in the outflow were significantly greater ($p < 0.05$) during fall than summer. This can be explained by the observation that there was reduced outflow and a longer retention time during summer compared to fall (see below).

In comparison to previous constructed treatment wetland studies (Gottschall *et al.* 2007; Newman *et al.* 1999), the ICW system has very high nutrient removal efficiencies, indicating its high treatment performance. The MRP concentrations were also efficiently removed in spring (95.7%), summer (90.7%), fall (94.9%), and winter (93.2%). The MRP outlet concentrations were significantly greater during fall than during summer because there was reduced outflow and a longer retention time during summer.

2.3.3.4 Flows

Mean monthly inflows and outflows from the ICW system during the monitoring period (2003–2007) were 180 ± 217.5 m³/month and 45 ± 132.3 m³/month, respectively. Although inflow to the ICW system was recorded usually between August and December, there was no corresponding outflow during most of this period, indicating that there was only a short discharge period to the receiving watercourse. The flows through the wetlands decreased during summer, and the ICW system was not discharging to the receiving watercourse.

Flow monitoring indicates that approx. 25% of the influent into the ICW system is subsequently discharged at the outlet as effluent. Based on flow and weather monitoring data, the breakups of inflow and outflow compositions are summarized in Figure 2.17. During dry periods, increased storage capacity was created within the ICW cells due to losses via evapotranspiration and infiltration. These processes provided additional storage buffer capacity for runoff within the ICW system prior to discharge during storm events.

The intermittent nature of the outflow and the great loss of partially treated runoff to the ground have a great impact on calculating and interpreting treatment performances. It follows that it is very difficult to accurately determine constituent masses. Therefore, calculations such as removal performances are based on con-

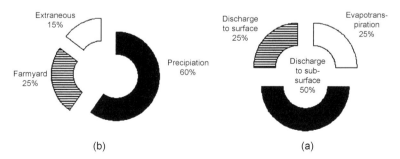

(b)   (a)

Figure 2.17 Composition of integrated constructed wetland: (a) inflow, and (b) outflow

stituent concentrations. Moreover, the lack of outflow data during drier in comparison to wetter periods is likely to result in less accurate interpretations for summer than winter data.

The microbial transformations within the wetland systems are a function of the available area for biofilm growth. In ICW systems, the dense vegetation stands and the associated litter provide large surface area for biofilm biomass and hence enhanced treatment capability. The biofilm entraps both organic and inorganic solids. The formation of biomass is greatest at the inlet of the wetland cell, where the organic loading is highest (Ragusa *et al.* 2004; Scholz 2006a), and decreases progressively as the wastewater passes through the system. The biofilm not only reduces the pollutant concentrations but also decreases the hydraulic conductivity by reducing the pore volume (Mustafa *et al.* 2009).

In addition to the biofilm formation, the organic matter and humic substances, which develop rapidly within the wetland soils, increase the availability of carbon supporting denitrification and the biological feedback mechanisms that secure water retention. These processes result in flow impedance and, subsequently, self-sealing of the ICW cell. This leads to a progressive contaminant reduction within the discharge to the groundwater.

2.3.3.5 Receiving Stream Water Quality

The construction and commissioning of the ICW system led to the transformation of a non-point source to an identifiable point source of potentially polluted water (*i.e.*, if treatment is insufficient). The land owner has, therefore, the responsibility to maintain the system to uphold BMP. However, there are currently no discharge standards for the final effluent in Ireland.

The final effluent of the ICW is discharged into a small stream through the outlet from cell 4. At the discharge point, mean outlet ammonia–nitrogen concentrations were 0.37 ± 0.562 mg/l, while nitrate–nitrogen and MRP concentrations were 0.99 ± 1.812 mg/l and 0.94 ± 0.628 mg/l, respectively.

The final ICW cell 4 is adjacent to the stream and has the second largest relative reduction in MRP, which is, however, not statistically significant. The most important and statistically significant ($p < 0.05$) observation is the great removal capacity of the first cell, where most SS-containing nutrients settle out.

Most nitrate–nitrogen concentrations at the point of confluence between the outlet of the final ICW cell 4 and the receiving stream, and after this point, are less than the corresponding concentrations for the preceding sampling points measured upstream. Further downstream, some measured nutrient concentrations decrease, indicating that the stream has sufficient assimilative capacity to buffer the ICW outflow concentrations, except for sampling point 14. This can be explained by two factors: the presence of a field drain discharging water rich in nutrients and the narrowing of the stream at that point. However, the trend reversed further downstream because field drains were absent.

Overall, the monitored stream had a median ammonia–nitrogen concentration of 0.05 mg/l, nitrate–nitrogen concentrations of 0.134 mg/l, and MRP concentrations of 0.029 mg. Phosphorus, which is a limiting nutrient, is in compliance with the Irish Phosphorus Regulations (1998). The key nutrient indicator for Irish rivers is MRP, which should be less than 0.03 mg/l for the annual median concentration.

Moreover, the ICW effluent is variable and depends predominantly on the patterns of precipitation. Outflows from the system are highly variable (1.5 ± 4.41 m^3/d) because the inflow quality and quantity of the ICW system is predominantly a function of precipitation intensities and patterns. Furthermore, there is no outflow from the system during periods of very low inflows, indicating that there is only a short discharge period to the receiving stream. Discharge from the ICW system during periods of high precipitation generally occurs during periods when the buffering capacity of the receiving water is enhanced by an increased dilution ratio (Mustafa *et al.* 2009).

2.3.3.6 Groundwater Quality

The ICW site was constructed using *in situ* soils. The subsoil was reworked and used in lining the bed and banks of all cells to reduce the risk of infiltrating pollutants. As the contaminated farmyard runoff passes through the ICW system, the suspended matter settles on the soil surface and subsequently hinders infiltration of contaminants through the wetland cells (Kadlec and Knight 1996; Scholz 2006a; Wallace and Knight 2006).

The mean ammonia–nitrogen concentration of the well located up-gradient was 0.79 mg/l, which indicates pollution but not originating from the ICW system. The nitrate–nitrogen and MRP concentrations were 0.02 mg/l and 0.12 mg/l, respectively. The mean ammonia–nitrogen concentration for the down-gradient well was 0.50 mg/l, which is less in comparison to the other wells located up-gradient and between ICW cells. Other nutrients determined down-gradient were below the detection limit of 0.02 mg/l. It follows that ammonia–nitrogen from the ICW is not impacting negatively on groundwater.

This finding is confirmed by an assessment of the nutrient data for the two wells between cells 3 and 4. The wells located between the ICW cells have a relatively high ammonia–nitrogen concentration because their sampling points are most influenced by the short distance to the cell sediments, which are rich in ammonia–nitrogen. The nitrate concentrations are also well below internationally recommended thresholds of approx. 25 mg/l nitrate.

Generally, the presence of high phosphate concentrations in groundwater is an indication of a shallow subsoil depth and the presence of preferential flow paths through the subsoil. However, in this case, where approx. 50% of the ICW water discharges to groundwater, the mean nitrate–nitrogen and MRP concentrations in all the down-gradient monitoring wells were <0.03 mg/l, indicating that the groundwater in the vicinity of the ICW system is not polluted by infiltration of contaminants from the ICW cells receiving nutrient-enriched runoff.

The soil investigation results at the time of construction indicated the presence of clay in the substratum of the ICW. Furthermore, the presence of hydrogel formed on the detritus confirms the low mobility of nutrients in subsoil and show the important role that subsoil plays in attenuating the pollutants.

Furthermore, there are many biogeochemical processes that play essential roles in impeding the infiltration of pollutants to the groundwater. For example, the ICW vegetation provides a large surface area to support microbial biofilms (Wallace and Knight 2006). Detritus provides a carbon source to microbes for denitrification and assists in the long-term sequestration of phosphorus (Wallace and Knight 2006). The production of methane during anaerobic metabolism inhibits the loss of water through its capillary-pore structures (Kellner *et al.* 2004; Tokida *et al.* 2005).

Virtually all ICW cells remove nutrients successfully. Nutrient enrichment of groundwater beneath the ICW system is of great concern, because high concentrations may have a direct effect on sensitive receptors (EPA 2002). The transportation and attenuation of pollutants in ICW cells depend predominantly on wetland soil, physical impedance and underlying geological formations; *e.g.*, attenuation of ammonia during migration in the sub-surface is known to occur (Erskine 2000). The key processes are sorption (cation exchange) and biological degradation control of ammonium in sub-surfaces (Buss *et al.* 2004). Concerning wetland peat, methane bubbles originating from microbial anaerobic processes lower the hydraulic conductivity (Kellner *et al.* 2004). The biofilm matrix formed on detritus acts like a hydrogel, which withstands changes in fluid shear stresses (Harrison *et al.* 2005).

2.3.3.7 Nitrogen Transformations Within the Sediment

In comparison to ammonia oxidizers, denitrifiers were more abundant in most of the collected sediment samples. Since the nitrate concentrations within the ICW systems were low, it is likely that oxygen and nitrate serve as electron acceptors in the lower layer of the wetland cells (Eriksson and Weisner 1996). This might have promoted the growth of denitrifying bacteria.

Denitrifying bacteria were detected in each cell of the ICW system, while ammonia-oxidizing bacteria were identified in the first, second, and third cells. Denitrifying bacteria were estimated to be present in higher numbers than ammonia-oxidizing bacteria.

The organic material present within the ICW cells also had an indirect impact on the bacterial community. For example, the litter on top of the sediment is likely to have limited the diffusion of oxygen to the lower sediment layers, creating anoxic conditions, and, hence, making conditions favorable for denitrification. This possible process has been described previously by Bastviken *et al.* (2005) for a comparable system. Most denitrifiers are heterotrophs, and the supply of organic carbon by macrophytes may have raised the overall heterotrophic activity, leading to the consumption of oxygen (Souza *et al.* 2008). Thus, it is likely that oxygen availability within the sediment was reduced, and subsequently denitrification was supported (Bastviken *et al.* 2005).

There was a reduced availability of organic matter at the bottom of ICW cell 1, where the greatest biofilm formation took place and the maximum biodegradation of organic matter occurred (*e.g.*, >85% BOD reduction). This led to decreased numbers of heterotrophic bacteria and consequently created conditions that favored the growth of ammonia-oxidizing bacteria.

The high number of denitrifying bacteria was linked to the accumulation of plant detritus over the years. The overall heterotrophic activity was increased by the supply of sufficient organic matter. This led to the consumption and subsequent reduction of oxygen within the sediment, thus supporting denitrification (Nielsen *et al.* 1990).

2.3.3.8 Integrated Constructed Wetland Sizing for Nutrient Reduction

Most ICW systems have a multicellular configuration with a minimum number of four cells. The systems operate as a set of sequential containment and treatment structures that intercept and control contaminants. The first cell receives the most contaminated influent. It is lined with organic matter. The loss of water through infiltration to the ground is kept low by the dense vegetation and associated litter with biofilms, while the subsequent wetland cells receive less contaminated water (Carroll *et al.* 2005).

A regression analysis for the 13 ICW located in the Annestown Stream watershed suggested that for sufficient MRP removal, a good correlation coefficient between the effective ICW area and the farmyard area can be determined (Scholz *et al.* 2007). The required effective ICW area for effective MRP removal should be at least 1.3 times the farmyard area. A factor of two is, however, recommended in practice to be on the safe side.

According to this design criterion, a farmyard area of $5000\,m^2$ required an effective ICW area for sufficient MRP removal of at least $6500\,m^2$. The actual ICW area was, however, 1.5 times the farmyard area, which is sufficiently large.

Most of the nutrient reductions occur in the initial cells. The effluent should stay as long as possible in the ICW system to allow for the development of mature ICW ecosystems effective in removing pollutants. This is achieved by allowing the pollutant plume to spread as slowly as possible throughout flatly designed ICW cells. Further details on sustainable wetland design have been published by Carty *et al.* (2008).

2.3.3.9 Overall Catchment Characteristics

Figures 2.18 to 2.21 show major overall catchment characteristics. The valley is dominated by pasture (Figure 2.18) used for cattle grazing. The runoff is usually higher in the upland areas, as indicated in Figure 2.19. Nitrogen loading rates are higher in the lowland areas (Figure 2.20). In comparison, no clear trend is visible for phosphorus (Figure 2.21).

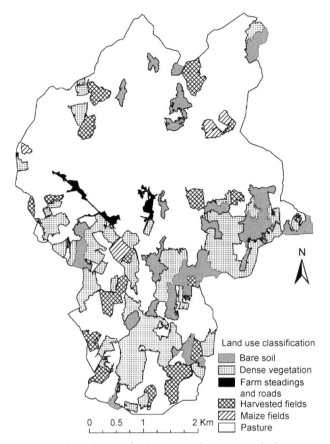

Figure 2.18 Land use in the Annestown Stream watershed

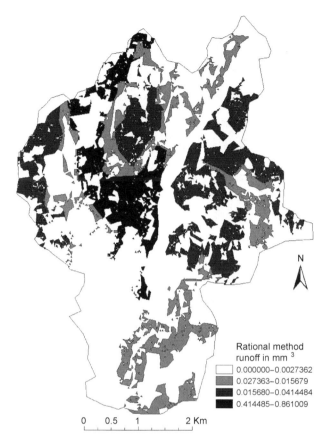

Rational method
runoff in mm ^{3}

☐ 0.000000–0.0027362
▨ 0.027363–0.015679
■ 0.015680–0.0414484
■ 0.414485–0.861009

0   0.5   1   2 Km

Figure 2.19 Annual runoff generated in the Annestown Stream watershed

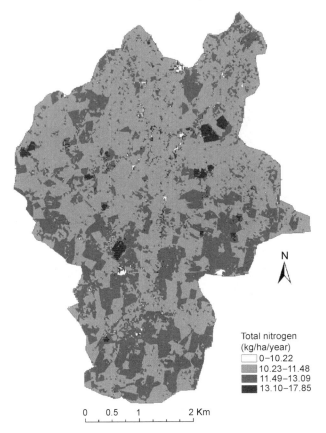

Total nitrogen
(kg/ha/year)
☐ 0–10.22
▨ 10.23–11.48
▨ 11.49–13.09
■ 13.10–17.85

0   0.5   1   2 Km

Figure 2.20 Total nitrogen loading rate calculated by the rational method for the Annestown Stream watershed

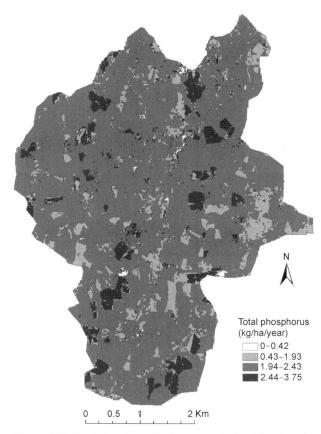

Figure 2.21 Total phosphorus loading rate calculated by the rational method for the Annestown Stream watershed

2.3.3.10 Soft Criteria

Landscape fit is one of the key objectives of the ICW concept (Harrington *et al.* 2005, 2007; Scholz *et al.* 2007b). The case study ICW system was therefore designed by utilizing the natural contours and features of the landscape. The utilization of natural contours resulted in reduced construction costs, leading to a cost-effective design of the ICW. Costs associated with cut and fill volumes were reduced by this technique. The shape of the valley and the crest and slope of hills were taken into account during the integration of the ICW site into the landscape.

Curvilinear shapes of cells that follow the natural contours were utilized to support the even distribution of flows within the system. The growth of vegetation in the cells of this ICW has assisted in the creation of a more natural look and a good fit with the surrounding landscape.

Furthermore, the ultimate aim of an ICW is to ensure that the water quality is improved and that the final result enhances the surrounding landscape and contributes to an improvement in the local biodiversity, as discussed by Carty *et al.* (2008). For example, the common newt (*Triturus vulgaris*) and invertebrates that are important indicators for 'good' water quality became abundant. Furthermore, *Dytiscus circumflexus*, a rare water beetle, has been identified along with mayflies and dragonflies within the case study ICW site boundaries. The ICW system has also attracted birds including the grey heron (*Ardea cinerea*) and mallard (*Anas platyrhynchos*). However, a quantitative assessment of the biodiversity was beyond the scope of this sub-section.

2.3.4 *Conclusions*

The studied ICW system reduced contaminant concentrations present in the farmyard runoff. Nutrients including ammonia–nitrogen and MRP were effectively reduced even after 7 years of operation. Hence, the structure acts as a sink (not source) for nutrients. Furthermore, the BOD, COD, SS, total coliforms, and *E. coli* values were also reduced within the ICW system.

Denitrifying bacteria were counted in greater numbers than ammonia-oxidizing bacteria within the sediments of the ICW cells. Thus, the benthos of this system supported denitrification and, consequently, made it a sink for nutrients.

The groundwater and surface water monitoring results indicated that the ICW system had not polluted the groundwater or degraded the water quality of the receiving watercourse. Nutrients including nitrate–nitrogen within receiving water bodies were lower downstream than upstream of the ICW system. However, the case study system is highly complex, and it is therefore difficult to define reliable nutrient balances.

The long-term assessment demonstrates that the example ICW system can be considered an effective and sustainable wastewater treatment option for farmyard runoff rich in nutrients. Dense stands of wetland plants contributed to very high evapotranspiration rates, which reduced the likelihood of floods further downstream during the summer by providing additional storage volume. Furthermore, the natural design enhanced the local biodiversity and improved landscape aesthetics.

2.4 Integrated Constructed Wetlands for Treating Swine Wastewater

2.4.1 Introduction and Agricultural Practice

Constructed wetlands have become widespread as an alternative and cost-effective method of treating wastewaters including agriculture-related wastewater streams (Harrington et al. 2005; Kivaisi 2001; Scholz and Lee 2005; Sievers 1997; Stone et al. 2002). As opposed to traditional mechanical and chemical treatment techniques such as sludge treatment plants or trickling filters, constructed wetlands implement only processes that occur naturally. The effective treatment of swine wastewaters (also called pig slurry in Europe) has been studied in greater depth over the past two decades, with more and more processes being examined and scrutinized.

The majority of research originated in the USA, where government agency guidelines were published for the construction and operation of constructed wetlands for various applications, but predominantly for domestic wastewater and storm water treatment (USEPA 1988, 2000). However, with respect to the complex nature of diffuse agricultural pollution, these guidelines were found to be inadequate (Sievers 1997; Stone et al. 2002, 2004) and resulted in mediocre overall treatment results due to the absence of a pretreatment stage promoting sedimentation or nitrification. Particularly livestock waste and wastewater are difficult to treat sufficiently by constructed wetlands due to their high concentrations of pollutants including suspended solids, nutrients, and bacteria (He et al. 2006). Moreover, an increasing number of livestock on farms and high production rates result in very high volumes of wastewater that prove to be a challenge to constructed treatment wetlands.

In the European Union, the Nitrates Directive 91/676/EEC (EEC 1991) has put pressure on piggeries due to the restrictions placed on land spreading of corresponding wastewaters at certain times of the year. Considering that suitable farmland for spreading is limited and other landowners are often unable to spread piggery waste as fertilizer, expensive and high-maintenance equipment is often purchased that can put great financial burdens on piggeries and farmers. The holistic approach of using constructed wetlands for the treatment of wastewater is therefore rapidly becoming a far more appealing approach to many farmers. However, constructed wetland acceptance will be hindered without greater understanding of the processes involved and guidance to their construction, operation, and maintenance.

The initially widely used guidelines from the USEPA (1988, 2000) were sufficient for certain types of wastewaters. Long and narrow systems often yielded respectable results despite high flow velocity measurements. This basic design has progressed to more sophisticated wetland system designs adopted to cold, temperate, and tropical climates (Kantawanichkul 2003; Puustinen 2005). Different designs include vertical-flow, horizontal surface-flow, and sub-surface-flow systems (Scholz 2006a).

The basis for the use of constructed wetlands for wastewater treatment is well established and its 'fine-tuning' for the use of intense agricultural wastes is ongoing. However, some systems such as ICW have produced exceptionally positive treatment results for farmyard runoff and domestic wastewater (Scholz *et al.* 2007) and are therefore being considered also for swine wastewater treatment.

The aim of this review is to assess the current level of knowledge and understanding with respect to the treatment of wastewater high in ammonia–nitrogen with constructed wetlands. The review will focus on swine wastewater treatment representing a 'worst-case scenario' for wetland designers. The reasoning for further research needs incorporating multidisciplinary and holistic views is also outlined.

2.4.2 *International Design Guidelines: Global Scenario*

2.4.2.1 American Guidelines

Since the early 1980s, a wide range of constructed wetland publications originating from the USA have been published. These include guidance manuals, case studies, bibliographies, assessments of technologies involved (both free surface water wetlands and sub-surface-flow wetlands), and the availability of a description and performance database, the so-called North America Treatment Wetland Database, which is based on constructed wetlands currently in use (Knight *et al.* 2000).

The US Environmental Protection Agency (USEPA) released a manual in 1988 for the design of constructed wetlands for the treatment of municipal wastewaters (USEPA 1988). The design parameters presented in the manual are based on a large number of empirical formulae and led to what is known as the "rational method" (Sievers 1997), which predicts wetland area based upon desired organic matter (5-d biochemical oxygen demand) removal.

The specific processes taking place in constructed wetlands are similar to those that occur in conventional treatment systems including treatment methods designed to promote denitrification. The use of emergent vegetation is crucial in both sub-surface and free surface water systems. The emergent vegetation controls water flow and provides the main source of treatment through the plants' rhizosphere where most treatment processes take place.

The USEPA (1988) design manual recommended the use of constructed wetlands for the treatment of acid mine drainage and storm water, and further suggested that constructed wetlands could also enhance existing predominantly natural wetlands. The manual acknowledged the considerable costs of more 'traditional' treatment systems and emphasized that constructed wetlands are a cost-effective alternative that are based on natural processes occurring in natural wetlands. The construction of wetlands in areas where they did not previously exist highlights that there is less regulation with regard to constructed wetlands than in comparison to natural wetlands. These and similar aspects about the ease of implementation and construction of constructed wetlands have been the driving factors behind the development of constructed wetlands for wastewater treatment since the late 1970s. Moreover, their performance, especially when compared to industrial methods, is a crucial factor promoting their increase of acceptance (Scholz 2006a).

The early design manual (USEPA 1988) was followed up by a series of guidance documents and manuals in the USA and elsewhere (Carty *et al.* 2008; Conte *et al.* 2001; USDA 1991; USEPA 2000). Similar guidelines were released by the United States Department of Agriculture under the Natural Resource Conservation Service (USDA 1991). This handbook recommends constructed wetlands for the treatment of agricultural and domestic wastewater, mine drainage, and storm water. Similar to the USEPA (1988) manual, it also relies on the constructed wetland design based predominantly on biochemical oxygen demand concentrations, but expanded the design parameters by including ammonia–nitrogen as the new key design factor and recommended the implementation of a minimum 12-d residence time for the influent to be treated.

Both manuals are similar, but have led to mixed results. Some reports have shown that nutrient concentrations were significantly higher than expected (Sievers 1997) when using the USEPA (1988) guideline. In comparison, others have shown that results are directly in line with the expected effluent quality using the USDA (1991) guideline (Stone *et al.* 2002).

The manual "Constructed Wetland Treatment of Municipal Wastewaters" (USEPA 2000) is recommended for the design of constructed wetlands as a functional part of a wastewater management strategy, and not as a guidance manual for a "specific regulatory program." It therefore gives in-depth guidance for both free surface-flow wetlands and vegetated submerged beds.

The USEPA (2000) manual describes constructed wetlands and the associated terminology, applications, and apparent misconceptions in a manner that is easy to understand. It allows for a wide array of people to comprehend the design, application, and implementation of constructed wetlands. However, its attempt to clarify certain aspects of constructed wetland design is confusing. On one hand, it explains that constructed wetlands are well defined by equations based upon a wide range of literature summarizing good research, but it states on the other hand that they cannot be designed purely by such equations. The document then gives example designs using these equations for the design of constructed wetlands for municipal wastewater treatment without outlining other design criteria in detail.

What has been consistent in the implementation of constructed wetlands is the focus on phosphorus treatment. It had been shown on several occasions that initial treatment was quite effective, but that after a relatively short period of time the reduction levels diminished considerably (Scholz 2006a; Sievers 1997; Stone *et al.* 2002). As phosphorus is deemed very important with regard to water quality, several studies recommend constructed wetlands only as a method of pretreatment when based upon the USEPA (1988) and USDA (1991) manuals.

There is little or no information in early guidance documents with regard to the use of constructed wetlands for the treatment of piggery wastewater. The USEPA online database does include some information for constructed wetlands that are or were being used for the treatment of piggery wastewater, but the information is not comprehensive. Usually, only the number of cells involved in the wetland systems themselves is stated. The USEPA and NRCS documents have both been implemented for the treatment of piggery wastewater and have shown inconclusive findings (Sievers 1997; Stone *et al.* 2002).

Sievers (1997) implemented the USEPA (1988) guidelines for four constructed wetland cells, which were run in parallel; two were surface-flow wetlands and two were free surface water wetlands. Only one cell of each wetland system type received wastewater from a primary anaerobic lagoon, while the remaining cells received wastewater from a secondary lagoon. The results from these cells were highly variable and most likely due to seasonal changes. The inability of the constructed wetland to consistently achieve a biochemical oxygen demand concentration of 30 mg/l or less was presumed to be a result of the wetlands' maintaining anaerobic conditions. Over a 2-year study period, no nitrate was detected in the constructed wetland system, thus suggesting insignificant nitrification. Phosphorus removal was also very low. The primary removal method was adsorption on soil. Despite low overall removal rates, the cells receiving wastewater from the primary lagoon provided relatively good treatment, most likely due to a longer retention time.

Stone *et al.* (2002) used the Natural Resources Conservation Service presumptive design method. This involved four wetland cells (*i.e.*, two cells connected in series). These cells had a length-to-width ratio of 9:1, and both were planted with either bulrushes or cattails (*Typhaceae*). Over a 6-year period, the macrophytes were examined while receiving effluent from an anaerobic lagoon, which was mixed with freshwater prior to entering the wetland.

The wetlands were found to be relatively effective at the removal of nitrogen with mean total nitrogen and ammonia–nitrogen reductions of approx. 85% each over the duration of the study period. At lower rates, the total phosphorus removal rate was approx. 88%. Total phosphorus removal was reduced to 25% and 38% for the two systems, if operated under higher loading rates. Stone *et al.* (2002) suggested that pretreatment would be required for the effective removal of phosphorus when high loading rates were applied. The constructed wetland sizing that was applied to these systems was dependent upon the loading rates and the concentrations of the wastewater being treated. This was found to result in a constructed wetland that was only slightly bigger (approx. 5%) than a similar system designed according to Natural Resources Conservation Service guidelines (see above).

The American guidance manuals (USDA 1991; USEPA 1988, 2000) have been relied upon for use in a wide range of applications with regard to various wastewaters. However, their usefulness regarding the treatment of swine wastewater was seen to be limited. Most constructed wetland designs lack the treatment capability to reach set standards for swine wastewater without pretreatment (Stone et al. 2002; Sievers et al. 1997). The manuals, while trying to be robust and highly descriptive, are overburdened with technical data and specific empirical formulae that are supposed to be applied to a treatment method that is extremely complex in both the reactions that are taking place in the constructed wetland as well as the external influences such as climatic conditions. These circumstances are stated in the more recent USEPA (2000) guidance manual for municipal wastewaters and described as a misconception.

Changes in the retention time impact the effectiveness of constructed treatment wetlands. The use of the Natural Resources Conservation Service method requires a minimum retention time of 12 d (Stone et al. 2002). Sievers (1997) noted improved performances in cells with longer retention times. This observation is taken into account in the more recent USEPA (2000) guidance manual. Earlier versions recommended between 6 and 7 d retention time.

The American guidelines for municipal water treatment are extensive and provide a large amount of information on all aspects of wetland design. They are recommended for industries and communities that can sustain the "expertise required to maintain them" (USEPA 2000). While being very technical in places, they do provide insight into all the major design aspects. However, there are no specific guidelines with regard to the implementation of constructed wetlands used for the treatment of agricultural wastewater. Therefore, municipal guidelines have been implemented by researchers and operators to examine their potential for use in swine wastewater treatment. However, these designs often fell short of the expected results (Cronk 1996).

2.4.2.2 Other Guidelines

Similar to what has been published in the USA, many other countries have their own guidance manuals for the use of constructed wetlands for the treatment of various wastewaters. Early guidelines were adapted for semi-arid (Australia), boreal (Finland), tropical (Thailand), Mediterranean (Italy), and temperate (Ireland, England, and Scotland) climates as discussed elsewhere (Carty et al. 2008; Conte et al. 2001; Puustinen and Jormola 2005).

The guidance manuals are used predominantly for the design and construction of wetlands for the treatment of municipal wastewater as well as storm water (Stone et al. 2002). Some of these guidelines are relatively new (Conte et al. 2001; Puustinen and Jormola 2005) in comparison to some that have been researched and implemented in other regions, namely Australia. A large proportion of the material that is used in the USA is based on data from Australia (USEPA 1988, 2000). Australia has been using constructed wetlands for several decades for the treatment of storm water in urban areas. Indeed, similar to the guidance material in the USA, other manuals and documents also primarily recommend constructed wetlands for the treatment of (pretreated) municipal wastewaters and storm water (Conte *et al.* 2001).

Several countries with areas characterized by a tropical climate such as China, Thailand, Malaysia, Brazil, and Australia have implemented constructed wetlands for wastewater treatment (He *et al.* 2006; Kantawanichkul 2003; Li *et al.* 2008; Sezerino *et al.* 2003). The functionality of constructed wetlands and their efficiency and size is influenced by the significantly higher temperature of their surroundings. Tropical constructed wetlands, like their temperate counterparts in Europe and the USA, are used predominantly for storm water and wastewater management (Kantawanichkul and Somprasert 2005), but are also being implemented for the treatment of agricultural waste and wastewater. The outline designs that are in use vary very little from those in most other regions, and the main differences are in wetland system operation. Specific guidance manuals have not been developed, and notes refer to the same design types as those used elsewhere. However, a combination of various types of constructed wetlands has been used in some case studies where space was limited (Kantawanichkul 2003, 2005).

Due to the high robustness of constructed wetlands and their wide treatment applications, they can even be found in harsh boreal climates that have extreme weather conditions. For example, the use of constructed wetlands in Finland for agricultural wastewater treatment is relatively new in comparison to other countries (Puustinen and Jormola 2005). Finnish designers rely heavily on optimizing the hydraulic efficiency, preventing flows that are linear, and preferring diffuse flows instead.

Constructed wetlands in Finland, for example, are designed and built with a broader spectrum of use than just a single purpose such as municipal or agricultural wastewater treatment. They are designed for floodwater mitigation, wastewater treatment, habitat creation, biodiversity support, and esthetic functionality (Puustinen and Jormola 2005). The primary Finnish research site is the Hovi constructed wetland in southern Finland (Puustinen and Jormola 2005). In the material documenting this wetland, it states that the size of the constructed wetland with regard to the catchment area is critical. These data are referenced against data from constructed wetlands in the USA and compared with data from Finland.

The high number of studies performed on this single wetland, which was established in 1998, has resulted in two manuals (Puustinen and Jormola 2005) provided by the Finnish Environment Institute. One manual gives an overview of the implementation of constructed wetlands, their role, and general information, and the other provides specific design guidelines.

2.4.2.3 Recent Innovations

The application of constructed wetlands for the treatment of agricultural waste and wastewater is not a new concept, but is one that has been refined since the construction of wetlands became viable for this specific application (Dunne *et al.* 2005a; Hunt *et al.* 2002, 2004; Knight *et al.* 2000; Sievers 1997). There have been several recent innovations regarding the design, operation, and management of constructed wetlands. For example, different types of vertical-flow wetlands are in use (He *et al.* 2006; Kantawanichkul *et al.* 2001; Molle *et al.* 2008; Sezerino *et al.* 2003). Some of these systems are common for swine wastewater treatment due to their increased capacity to remove organic matter and ammonia–nitrogen (He *et al.* 2006).

Different methods of introducing wastewater into constructed wetlands (*e.g.*, continuous, batch, and "tidal" (Lee *et al.* 2006) flow operation) and "antisized reed bed systems" (Zhao *et al.* 2004), where, in contrast to normal vertical-flow systems with finer material at the top of the system, the coarser material is located at the top instead, have been successfully operated (Lee *et al.* 2006; Scholz 2006a; Scholz and Xu 2001; Zhao *et al.* 2004). Zhao *et al.* (2004) examined the improved capability of such systems to deal with the clogging of the bedding material, as well as avoiding the possible detrimental effects of clogging on functionality and sustainability. The study found very little difference in the actual removal capacity of nutrients in the systems, but the accumulation of material in the set-ups was significantly slower in the antisized systems. In situations where vertical-flow wetlands are used, this would lead to the improved sustainability of the constructed wetland, especially in more remote areas where access to skilled personnel would be more difficult (USEPA 2000).

Another, more encompassing, approach has been undertaken in Ireland with regard to the overall design, construction, and management of wetlands. The ICW concept has been developed for the design of wetlands for agriculture including swine wastewater interception, and the design takes into account rather holistic variables before construction (Harrington *et al.* 2005). These additional variables are primarily of a social, economic, and environmental nature. The ICW approach also integrates wetlands into the surrounding landscape to give a more natural appearance of the overall structure (Scholz *et al.* 2007b).

Carty *et al.* (2008) have published the scientific justification for the Farm Constructed Wetland Design Manual for Scotland and Northern Ireland. This document addresses an international audience interested in applying wetland systems in the wider agricultural context. Most FCW combine farm wastewater treatment with landscape and biodiversity enhancements and are a specific application and class of ICW that have wider applications in the treatment of other wastewater types such as domestic sewage (Scholz *et al.* 2007b). Carty *et al.* (2008) discuss universal design, construction, planting, maintenance, and operation issues relevant specifically for FCW, including wetland systems treating piggery wastewater in a temperate climate, but highlight also catchment-specific requirements to protect the environment.

Furthermore, the design suggestions by Healy *et al.* (2007) complement the guidelines proposed by Scholz *et al.* (2007b) and Carty *et al.* (2008). Healy *et al.* (2007) discuss the performance and design criteria of constructed wetlands for the treatment of domestic and agricultural wastewater, and sand filters for the treatment of domestic wastewater. It also proposes sand filtration as an alternative treatment mechanism for agricultural wastewater and suggests corresponding design guidelines.

Chen *et al.* (2008) examined modified free water surface-flow constructed wetlands to polish treated swine wastewater. The assessed treatment process (*i.e.*, conventional three-stage treatment scheme followed by a modified free water surface wetland (with or without plants) with a 2-d hydraulic retention time) was a promising option to meet Chinese swine wastewater discharge restrictions (COD: 600 mg/l, BOD: 80 mg/l, SS: 150 mg/l).

Most recently, Zhang *et al.* (2009) proposed the application of self-organizing map models as prediction tools for the performance of wetland-based agroecosystems for the treatment of agricultural wastewater to protect receiving watercourses. By utilizing the self-organizing map model, the expensive biochemical oxygen demand outflow concentrations, which are time consuming to measure, were predicted well by other inexpensive variables, which were quicker and easier to measure. This novel approach allows for the real-time control of the outflow water quality of wetland systems and potentially also of other treatment system applications.

2.4.3   Operations

2.4.3.1   Loading and Flow Rates

The operation of standard constructed wetlands is relatively uniform, regardless of the flow type (*e.g.*, surface flow, sub-surface flow, or vertical flow). The influent enters the wetland system and flows through one or a series of wetland cells, which are usually heavily vegetated. Where the differences lie are the various aspects with regard to specific design.

The design of constructed wetlands is therefore of great importance to ensure that they perform well and are suited to the location and to the treatment role that they are supposed to fulfill. Comprehensive research into the most appropriate loading rates, input concentrations, and flow rates has been undertaken (Kantawanichkul *et al.* 2001; Knight *et al.* 2000; Lee *et al.* 2004; Szogi and Hunt 2001).

The North American Treatment Wetland Database contains design data for over 400 constructed wetlands, both study- and full-scale systems. A study sponsored by the Gulf of Mexico Program established a similar database, the Livestock Wastewater Treatment Database (Knight *et al.* 2000). The North American Treatment Wetland Database stores information on all types of treatment wetlands, whereas the Livestock Wastewater Treatment Database stores information specifically for livestock-related wastewaters. These two databases were combined due to their similarity in design to form the North American Database version 2.0 (Knight *et al.* 2000).

The loading and flow rates in constructed wetlands vary greatly. The Livestock Wastewater Treatment Database recorded data from 68 sites in North America. The mean hydraulic loading rate was 4.7 cm/d. Mean system flows of 10 m^3/d have been calculated. Most of the systems used for livestock wastewater treatment are of small size with a mean area of only 0.6 ha. Constructed wetlands treating swine wastewater are slightly larger with mean areas of 1 ha. Knight *et al.* (2000) concluded that concentration removal rates were a function of the inlet concentration and the hydraulic loading rates.

The USEPA (1988) guidelines recommend loading rates of 112 kg BOD_5/ha/d and flow rates of 200 m^3/ha/d. The Livestock Wastewater Treatment Database stated mean flow rate of 10 m^3/ha/d is significantly lower than this, but the majority of the sites examined in the livestock database were research-based systems of relatively small size.

The loading rates that are used in constructed wetlands vary greatly with respect to their intended use. In dairy wastewater treatment, designers have applied very light loading rates (3.6 g/m^2/d; Dunne *et al.* 2005a), heavy rates (Lee *et al.* 2004; USEPA 1988), tidal-flow-governed rates (Zhao *et al.* 2004), and changing rates (Lee *et al.* 2004). Several studies in the tropics with high hydraulic loading rates in constructed wetlands treating swine wastewater have shown that removal rates decrease with significant increases in loading rates (Kantawanichkul and Somprasert 2005; Kantawanichkul *et al.* 2003).

2.4.3.2 Water Depth

The water depths in constructed wetlands are highly variable. The USEPA (1988) guidelines recommend that the variable depth should be part of the design equations that are used to determine the sizing of the corresponding constructed wetland cells. Wetlands range from shallow systems with a depth of approx. 0.25 m (Harrington and Ryder 2002; Harrington *et al.* 2005) to deep ones of 1.2 m (USEPA 1988). A technical report by the USEPA (1999) recorded water depths of between 0.1 and 1.5 m. The depth of water is often the key parameter to nitrogen control in wastewater treatment. Shallow water depths are associated with the highest ammonia–nitrogen diffusion and nitrogen losses (Szogi and Hunt 2001).

The water depth also has an important impact on the growth of macrophytes that are used in constructed wetland systems. Depending on the species that are planted, greater water depth can inhibit macrophyte growth and colonization of the wetland cells. Some genera of macrophytes are capable of coping with deep water levels (Clarke and Baldwin 2002; USEPA 1988), but others are not and therefore require shallower water for most of their growing season. High water levels would result in wetlands being more susceptible to the effects of high nutrient concentrations and may lead to the death of more sensitive species. Shallower systems also help to increase nitrification by increasing the aerobic conditions present in the cells (Carty *et al.* 2008; Scholz 2006a).

2.4.3.3 Pretreatment of Wastewater

Pretreatment is often important to achieve a better and more effective treatment of wastewaters (Cronk 1996; Hunt and Poach, 2001). Early guidelines (USEPA 1988) recommended the use of conventional pretreatment units such as sedimentation basins to reduce suspended solids and BOD concentrations and suggested the addition of chemicals to remove phosphorus.

Constructed wetlands are often viewed as long-term solutions to wastewater treatment. They are low-tech and low-cost and have a design life of usually more than 25 years. However, their longevity can be impeded by the accumulation of detritus and sediment build-up. The removal of solids in a pretreatment stage has therefore been recommended for many years (Hunt and Poach 2001; Cronk 1996; USEPA 1988). Some studies have suggested that some constructed wetland types themselves can be used as a pretreatment step, being part of a sustainable drainage system, also called BMP in the USA (Cook *et al.* 1996).

Furthermore, the dilution of wastewaters to improve treatment is a common practice in constructed wetland operation (Harrington *et al.* 2005). This method is often just a simple addition of water to wastewater. However, heavily polluted wastewater can also be diluted by less contaminated wastewaters such as roof and yard runoff. The dilution of wastewaters is important in promoting good nutrient removal within constructed wetlands. Moreover, if the organic loading rates are excessive, this can result in decreased removal performances (Kantawanichkul *et al.* 2003) and an increase in the risk of ammonia toxicity to some constructed wetland plants (Hunt *et al.* 2002, 2004).

Partial nitrification of wastewater prior to its treatment in a constructed wetland has also been examined as a means of affecting nitrogen removal and ammonia volatilization (Sommer *et al.* 1993; Poach *et al.* 2003). The most common method of additional nitrification of wastewaters is the recirculation of the wastewater itself or the use of partially nitrified lagoon wastewater, as shown by numerous previous researchers (He *et al.* 2006; Humenik *et al.* 1999; Hunt and Poach 2001; Kantawanichkul *et al.* 2001; Poach *et al.* 2003).

Pretreatment to achieve nitrification can reduce ammonia volatilization due to reduced concentrations of ammonia in the wastewater (Poach *et al.* 2003). As a result, these reduced concentrations help to minimize the potential risk of ammonia toxicity to wetland plants. The anaerobic conditions that exist in wetland soils (Scholz 2006a) limit the rate of nitrification, suggesting that denitrification in wetlands is nitrate-limited (Hunt and Poach 2001; Hunt *et al.* 2002, 2004).

Recirculation of wastewaters has been shown to have a positive effect on total nitrogen removal across the world. Kantawanichkul (2001) showed that the recirculation of effluent in a combined vertical-flow and horizontal-flow wetland system increased the total nitrogen removal rate from 71 to 85%. Humenik (1999) reported that nitrified lagoon water added to constructed wetland microcosms led to nitrogen removal rates that were four to five times that of non-nitrified liquid.

Pilot-scale plants in China (He *et al.* 2006) were used to test for various recirculation rates. Findings showed increased ammonia–nitrogen removal rates in comparison to non-recirculated effluent. However, the researchers reported no great increase in phosphorus removal. This study showed that the use of recirculation helped to form an oxide environment in the wetland.

Furthermore, the partially nitrified state of recirculated effluent has consistently shown its benefit to nitrogen removal in constructed wetland systems. Recirculation technologies decrease ammonia volatilization, which is not desirable due to it being an atmospheric pollutant (Sommer *et al.* 1993; Poach *et al.* 2004), and help to abate ammonia toxicity with regard to the most commonly used macrophytes.

2.4.4 Macrophytes and Rural Biodiversity

2.4.4.1 Macrophyte Types and Characteristics

The range of macrophytes that are planted in constructed wetlands is wide and varied. They are intrinsic to the use of constructed wetlands and play a vital role in nutrient removal (Brix 1994; Hunt and Poach 2001). The USEPA (1988) guidance manual refers to cattails, reeds, rushes, bulrush, and sedges, all of which have different ranges of pH tolerance. For example, cattails usually tolerate pH values of between 4 and 10, while other aquatic plants such as rushes and sedges have much narrower tolerance margins. The USDA (1991) guidebook on constructed wetlands lists several species that have been identified as being suitable for use in constructed wetland systems in North America. The guidebook also states that not all wetland plants are suitable for treatment systems, since they should be able to tolerate continuous flooding and exposure to high nutrient concentrations in the influent (USDA 1991). Tanner (1996) summarized a list of properties that wetland plants should have:

- ecological acceptability (no significant weed or disease risks or danger to the ecological or genetic integrity of surrounding natural ecosystems);
- tolerance of local climatic conditions, pests, and diseases;
- tolerance of pollutants and hypertrophic waterlogged conditions;
- ready propagation and rapid establishment, spread, and growth; and
- high pollutant removal capacity.

The principal functions that macrophytes provide are numerous: stabilization of the beds, provision of physical filtration, prevention of clogging of vertical systems, insulation against frost in winter, and provision of a large surface area for microbial communities, which are vital to successful wastewater treatment (Brix 1994; Scholz 2006a). In addition to supporting the treatment processes in constructed wetlands, macrophytes also serve highly underrated functions in traditional civil engineering design by promoting natural aesthetics and landscape integration.

Furthermore, planting of the most suitable and often native species is important in the ICW concept to improve the biodiversity of the vicinity around the structure (Harrington *et al.* 2005; Scholz *et al.* 2007b). The predominantly aquatic plants provide habitats for wildlife such as mammals, birds, and insects. The biodiversity of macroinvertebrates has been shown to be extremely high in certain ICW in

Ireland, *e.g.*, some wetland systems have up to 60% of the country's native species of aquatic macroinvertebrates present. The adaptation of wetland plants to live in anaerobic soils is important as their root structures provide aerobic areas that help to sustain nitrifying bacteria (Brix 1994). As well as providing oxygen for bacteria, they also provide oxygen to the anaerobic substrate and thus help to stimulate aerobic decomposition.

2.4.4.2 Toxicity Tolerance Thresholds

The toxicity tolerance thresholds and the corresponding uptake rates of pollutants by wetland plants have been researched (Brix 1994; Harrington 2005; Hill *et al.* 1997; Hubbard *et al.* 1999). However, these studies have usually examined the more common genera used in constructed wetlands (Brix 1994; Clarke and Baldwin 2002; Hubbard *et al.* 1999). For example, Clarke and Baldwin (2002) tested common species such as softrush (*Juncus effuses* L.), broadleaf arrowhead (*Sagittaria latifolia* Willd.), softstem bulrush (*Schoenoplectus tabernaemontani* C.C. Gmel.), lesser bulrush (*Typha angustifolia* L.), and common bulrush (*Typha latifolia* L.) at varying ammonia concentrations and water depths.

Other studies have assessed more genera growing in temperate or tropical climates (Tanner 1996). However, findings concerning the effect of ammonia on plants are not fully conclusive. The preference to pretreat wastewater prior to it entering the wetland system or the recycling of effluent suggests that the aquatic plants studied were most likely susceptible to ammonia toxicity, although some studies suggest that the plants are more tolerant than is commonly reported in the literature, stating that there was no apparent effect on some plants due to relatively high ammonia concentrations (Hill *et al.* 1997).

Comparisons between different plant species have been undertaken to examine their uptake rates (Hubbard *et al.* 1999; Poach *et al.* 2003; Tanner 1996). Brix (1994) reported on the uptake rates of common emergent, free-floating, and submerged plant species in wetlands. For example, bulrush (*Typha latifolia* L.) had an impressive nitrogen uptake rate for relatively small planted areas, but low phosphorus uptakes for considerably larger areas. The nutrients are, however, bound in the biomass but could be removed by harvesting.

With plants being an important integral part of constructed wetlands, attention has been paid to the opportunity to use them for additional purposes. Therefore, cash crops such as soybean and rice have been assessed in terms of their use in wastewater treatment (Humenick *et al.* 1999; Szogi *et al.* 2000, 2003, 2004).

These research studies indicate that such plants are able to grow in treatment wetlands receiving swine wastewater. The potential yield from such cash crops could make constructed wetlands that use these plants more attractive, particularly in developing countries. Constructed wetlands could be used for the treatment of wastewater and also to yield a steady food supply or income. This would be particularly beneficial for small-scale farmers, because they would be able to produce their own feed while treating their own wastewater at the same time.

Alternative methods of using aquatic plants are not limited to cash craps. The most common macrophytes planted on floating mats in anaerobic lagoons treating swine wastewater have been assessed by Hubbard *et al.* (2004). The nutrient uptake rates were relatively high. Less commonly used plant species native to certain regions have been examined as well. For example, vetivergrass (*Vetiveria zizanioides* Nash) was used in Thailand (Kantawanichkul *et al.* 2003). This grass was suitable for tropical hydraulic and organic loading rates.

2.4.5 Nutrients

2.4.5.1 Nutrient Transformation Processes

The primary objective of constructed wetlands is the removal of excessive concentrations of nutrients in wastewater (Lee *et al.* 2004; Poach *et al.* 2002; Prantner *et al.* 2001). The removal is supported by means of sedimentation, adsorption, organic matter accumulation, microbial assimilation, nitrification, denitrification, and ammonia volatilization (Brix 1994; Poach *et al.* 2003). However, denitrification is far more preferable to ammonia volatilization as a means of nitrogen reduction, considering that ammonia volatilization results in ammonia gas release, which is an atmospheric pollutant (Sommer *et al.* 1993; Poach *et al.* 2004). Denitrification, however, is limited in constructed wetlands by the availability of nitrate and nitrite (Hunt and Poach 2001; Hunt *et al.* 2002, 2004).

Denitrifying enzyme activity is correlated to areas where there are high amounts of nitrate and nitrite. Moreover, denitrifying enzyme activity has been shown to increase over time with the maturity of constructed wetlands, an increased rate of nitrogen application, and water depth (Hunt *et al.* 2002), which leads to increased ammonia volatilization. This is seen as a particular problem, especially with regard to swine wastewater and its high nutrient content.

The phosphorus cycle is dissimilar to both the carbon and nitrogen cycles in that it does not involve a series of oxidation-reduction reactions. In comparison, it is predominantly a sedimentary cycle (Van der Valk 2006), as shown in Figure 2.22.

The most common method of enhancing denitrification is to recirculate the wastewater or to add partially nitrified water (He *et al.* 2006; Kantawanichkul *et al.* 2001; Poach *et al.* 2003). This is done by recycling the effluent back into the system or by the addition of partially nitrified storage or lagoon water. It follows that denitrification is promoted by supplying the system with greater amounts of nitrate and nitrite throughout the treatment process, thereby reducing the risk of volatilization. However, a complete removal of the volatilization process is unrealistic.

Seasonal and temperature effects on denitrification have also been examined (Reddy *et al.* 2001; Trias *et al.* 2004). There is a moderate positive correlation between temperature and the removal rates in wastewaters. Trias *et al.* (2004)

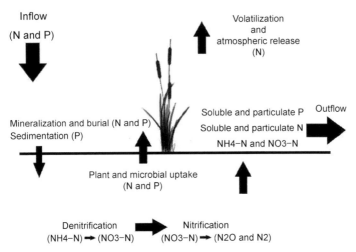

Figure 2.22 Simplified phosphorus and nitrogen cycles within integrated constructed wetlands treating swine wastewater

reported variable findings for swine wastewater treatment with respect to total suspended solids ranging from 77% at moderate temperatures to 42% during the warmest period.

Studies with marsh–pond–marsh designs have assessed differences in the amount of ammonia volatilization taking place in constructed wetlands (Poach *et al.* 2004; Reddy *et al.* 2001). While Reddy *et al.* (2001) saw rates of nitrogen removal similar to those of continuous marsh systems (often >70%) and medium phosphorus removal rates (30 to 45%), it was not observed how much of the nitrogen removal was associated with volatilization.

Poach *et al.* (2002) highlighted the fact that it was not known if volatilization of free ammonia governed nitrogen removal in constructed wetlands. Findings indicated that between 7 and 16% of the nitrogen removal was achieved by ammonia volatilization. This gave an indication as to how much nitrogen removal was being caused by volatilization, but also showed that it was not the principal nitrogen removal method.

Poach *et al.* (2004) used the marsh–pond–marsh design a further time to highlight the differences between the pond and marsh sections with regard to volatilization. The pond sections had significantly higher proportions (23 to 36%) of volatilization than the marsh areas (<12%). Volatilization was the dominant nitrogen removal mechanism in the pond sections (54 to 79%). It follows that marsh areas should be constructed within wetland systems used for the treatment of wastewaters from confined animal operations such as swine waste and wastewater. However, relatively low removal rates of nitrogen (30%) and phosphorus (8%) have been reported for long and narrow marsh–pond–marsh systems in the state of North Carolina, USA (Stone *et al.* 2004).

2.4.5.2 Phosphorus

Phosphorus is among the most difficult nutrients to remove from wastewater. This has created problems for many types of treatment systems such as constructed wetlands, where it is retained but often only temporarily (Pant *et al.* 2001; Reddy *et al.* 2001; Sievers 1997; Stone *et al.* 2002, 2004). Like nitrogen, phosphorus is partly removed in constructed wetlands by plant uptake, accretions of wetland soils, microbial immobilization, retention by root bed media, and precipitation in the water column (Reddy *et al.* 2001; Scholz 2006a).

Phosphorus removal rates in the literature are very diverse and subject to the boundary conditions of the individual case study. Some research studies with constructed wetlands have shown average phosphorus removal rates (He *et al.* 2006; Knight *et al.* 2000; Lee *et al.* 2004; Reddy *et al.* 2001; Sezerino *et al.* 2003). However, some other studies have reported very low rates (Shappell *et al.* 2007; Stone *et al.* 2004; Szogi *et al.* 2004), while a relatively large proportion of investigations have shown very high removal rates, usually >70% (Harrington *et al.* 2005; Hunt and Poach 2001; Prantner *et al.* 2001).

In wetland soils, phosphorus is present as soluble or insoluble, organic or inorganic complexes (Scholz and Lee 2005). The phosphorus cycle is not gaseous, as opposed to that of nitrogen; rather it is sedimentary and is retained in the wetland (Figure 2.22). Inorganic or mineralized organic phosphorus may be retained by oxyhydroxides of iron and aluminum in acidic soils and by calcium minerals in alkaline soils (Reddy and D'Angelo 1997; Scholz and Lee 2005). Soluble phosphorus can be removed by periphyton and subsequently by deposition of dead biomass on soil and detritus surfaces (Reddy and D'Angelo 1997).

In aerobic conditions, insoluble phosphates are precipitated with ferric iron, calcium, and aluminum (Drizo *et al.* 2002; Scholz and Lee 2005). This has led to experiments with various aggregates that are used as the primary substrate in constructed wetlands (Drizo *et al.* 2002; Scholz and Xu 2001). Most of these compounds readily bond to phosphorus and are seen as an effective method of increasing or sustaining phosphorus removal and retention within constructed wetlands.

The partly aerobic and partly anaerobic conditions within wetlands are important for phosphorus removal. In anaerobic soils and detritus, forms of organic phosphorus are relatively resistant to enzyme hydrolysis, which would otherwise release the phosphorus back to the bioavailable pool of nutrients (Reddy and D'Angelo 1997). In aerobic conditions, however, phosphorus is bound to organic matter and is incorporated into bacteria, algae, and macrophytes (Brix 1994; Reddy and D'Angelo 1997; Scholz and Lee 2002).

As phosphorus is retained in the constructed wetland itself or released in low concentrations as part of the wetland discharge, there has been concern among wetland critics with regard to the long-term ability of constructed wetlands to retain phosphorus effectively. Nevertheless, there seems to be an agreement that wetland size matters, *i.e.*, the larger the wetland, the better the long-term removal of phosphorus through adsorption and precipitation (Harrington *et al.* 2005; O' Sullivan *et al.* 1999; Pant *et al.* 2001). It follows that an appropriately sized wetland system has a long design life and its capacity to treat phosphorus and other nutrients can be greatly increased. Moreover, the design life can be enhanced by harvesting macrophytes at the end of the growing season. This maintenance method leads to the removal of large amounts of phosphorus captured within a relatively small amount of biomass.

2.4.6 Pathogens, Odor, and Human Health

The ability of wetlands to remove pathogens and bacteria is well documented. Wetlands are similar to biofilters, which are governed by the following processes (Brix 1994; Scholz 2006a):

- chemical: oxidation; adsorption; exposure to toxins released by microorganisms;
- biological: antibiosis; ingestion by nematodes, protozoans, and cladocera; lytic bacteria; bacteriophage attacks; natural decay; and
- physical: filtration; sedimentation; aggregation; ultraviolet radiation.

The removal of pathogens from swine wastewater is important for human health and safety reasons. Pathogens such as *Salmonella* may enter the human body via surface waters contaminated by runoff containing traces of swine wastewater, which has been land-spread. Considering the treatment of swine wastewater with constructed wetlands, very high removal rates of enteric microbes and pathogens such as *Salmonella*, faecal coliforms, and *E. coli* (Hill and Sobsey 2001) as well as *Giardia* and *Cryptosporidium* (Karim *et al.* 2004) have been reported.

Hill and Sobsey (2001) documented pathogen removal ratios of between 70 and 90% in surface-flow wetlands and between 80 and 99.99% in sub-surface-flow wetlands with varying organic loading rates. A study of the removal of enteric pathogens showed that there were greater occurrences of *Giardia* cysts and *Cryptosporidium* oocysts in the sediments than in the water column itself. This may be indicative of the greater removal rates with regard to pathogens in sub-surface-flow systems.

Odor removal in wetlands has become an important factor with regard to the construction of wetlands for agricultural wastewater treatment systems (Harrington *et al.* 2005). The capacity for well-vegetated systems to reduce malodors is important for their acceptance, particularly in populated areas. Wetland acceptance would result in their wider application in wastewater treatment, especially for municipal systems and rural farming communities.

The odor-reduction capability of livestock wastewater treatment wetlands has been primarily anecdotal (Wheeler *et al.* 2007). However, studies have been performed to quantify odor from constructed wetlands treating swine wastewater. Findings indicate that constructed wetlands are very capable of reducing malodor from the wastewater being treated, which is seen as an additional benefit (Wheeler *et al.* 2007).

2.4.7 Conclusions and Further Research Needs

Wetlands have been part of human society for hundreds of years, but it is only in the past several decades that they have been used with the specific intent and purpose of treating wastewater. Different wetland systems, regardless of their given title (*e.g.*, constructed wetlands, treatment wetlands, or ICW), have proven to be highly efficient at treating agricultural wastewaters. A wide range of studies on their operation and inner workings allowed for much better understanding of the biochemical processes that take place. This led to their application in the treatment of high strength ammonia–nitrogen polluted waters such as swine wastewater, which is perceived to be difficult to purify because of its high strength and ammonia content.

Due to their high efficiency and relatively low land use, cost, maintenance, and operation effort, wetland systems are ideal as an alternative to traditional wastewater treatment technologies for the treatment of swine wastewaters. The restrictions that are put upon piggery farmers, for example, by the European Union Nitrates Directive prevent farmers from using methods of wastewater disposal such as land spreading that are not sustainable. The use of constructed wetlands for the treatment of swine wastewater, either as its main treatment method or as a secondary or even tertiary treatment, has shown consistently high removal rates for nutrient and other pollutants as well as a reduction of distinctive malodor that is associated with swine waste.

The greater use of ICW should be considered as a standalone technique or in conjunction with traditional aerobic digesters or land-spreading methods to improve the overall treatment efficiency, reduce costs, and allow for a more holistic approach to wastewater treatment.

There is a further research need for more well-documented case studies demonstrating good practice in the treatment of ammonia-rich wastewater with ICW systems. The greatest challenges entail overcoming the toxicity thresholds for different plant species and avoiding groundwater contamination if no impermeable artificial liner is applied for economic reasons. Furthermore, there is the potential that novel analytical tools such as self-organizing map models may help to improve traditional constructed wetland guidelines and manuals by optimizing design calculations based on artificial intelligence.

2.5 Wetlands to Control Runoff from Wood Storage Sites

2.5.1 Introduction and Objectives

Handling of wood in different shapes and forms at small or industrial scales takes place in all regions with climatic conditions allowing trees to grow and people to reside. Wood is an important natural resource and is used all over the world as, for example, an energy source, building material, raw material for paper, furniture, and tools, and for art and decoration purposes. The environmental impact of logging, transport, storage, debarking, sawing, milling, chopping, and pulping of wood depends on the scale and location of the operation, tree species, handling methods, and preventative measures that have been chosen to reduce the corresponding impacts on the environment and receiving watercourses (Zenaitis *et al.* 2002).

The environmental impacts from sites handling wood include emissions of particles (Kauppinen *et al.* 2006) and volatile organic compounds from wood (Rice and Erich 2006; Welling *et al.* 2001), emissions from the use of energy fuels (Zelikoff *et al.* 2002), release of storm water that has been in contact with wood or irrigation water used to protect wood (Bailey *et al.* 1999; Woodhouse and Duff 2004), and emissions and spillages from machines and vehicles used at the site (Orban *et al.* 2002; WDOE 1995). Furthermore, noise (Kraus 1985), odor (Nicholson 1987), and light (Chalkias *et al.* 2006) emissions from wood handling sites are also seen as undesirable.

This section focuses on the impacts of wood storage and the corresponding mechanical wood handling processes on the receiving watercourses and the environment and assesses the techniques that could be used to reduce adverse impacts. The specific environmental effects of logging and transport (Lindholm and Berg 2005) are not discussed in this section. A detailed discussion on the environmental impacts of water emissions from pulp and paper mills (McMaster *et al.* 2006) is also excluded because the impact from these sites originates to a large extent from the extensive use of chemicals. However, a summary of experimental treatment techniques for organic material originating from wood in effluent from pulp and paper mills is included in this section because these methods can also be used for runoff treatment from other sites handling wood.

This sub-chapter summarizes the findings published in high-impact journal papers to enable further research into developing cost-effective and low-tech treatment methods to reduce the environmental impact of runoff from wood handling sites. The first sub-section presents the background and purpose of this section. The second sub-section describes the pollution potential that storage and different kinds of mechanical handling of wood have on the groundwater or the receiving watercourses in the vicinity of the handling sites. The third sub-section summarizes the methods that are currently applied to reduce the pollution. The section concludes with a brief discussion regarding the cost-effectiveness of the different methods for reducing the environmental impact of runoff, a summary of conclusions, and key recommendations for relevant future research.

The sub-chapter focuses on wetland treatment and soil infiltration because this kind of treatment can, given suitable ground conditions, be applied at sites of different sizes at a relatively low cost. Some more high-tech methods are also discussed briefly. The objectives are as follows:

- to summarize the reasons for pollution generation at wood handling sites;
- to characterize the runoff and its effect on the receiving watercourses;

- to summarize treatment methods including soil infiltration and wetland treatment; and
- to discuss treatments (used in recent experiments) for organic matter in pulp and paper mill wastewater.

2.5.2   Pollution Potential of Runoff from Wood Handling Sites

2.5.2.1   Reasons for Pollution Generation at Wood Handling Sites

Contaminated storm water runoff from log yards is generated when precipitation comes into contact with wood, woody debris, and equipment at outdoor wood sorting, processing, and storage facilities (Woodhouse and Duff 2004). Runoff can also be generated by applications of water for dust and fire control (Orban *et al.* 2002). In northern Europe and Canada, it is also common that stored logs are irrigated with water to protect them from cracking and biological attack (Webber and Gibbs 1996). The water keeps the moisture content in the sapwood above 50%, which prevents them from drying and being damaged by fungi and insects (Liukko and Elowsson 1999). There are techniques, such as climate-controlled sprinkling (Liukko 1997), that can be applied to minimize the amount of sprinkling water (Figure 2.23). The sprinkling intensities can be decreased with this method by between 31 and 97% over a 24-h period (Liukko and Elowson 1995). However, runoff is not completely eliminated because of necessary safety margins and maintenance water pressures in the sprinkling systems. A medium-sized log yard in central Sweden with climate-adapted sprinkling in a non-recirculating system uses approx. 100,000 m^3 water between May and September. During this period, the amount of log yard runoff may be as high as 70,000 m^3 (Jonsson 2004).

Figure 2.23 Log sprinkling resulting in large amounts of sprinkling water that runs off from a log yard

Soluble compounds and particles from bark and wood are taken up by the water and become part of the site runoff (Bailey *et al.* 1999; Woodhouse and Duff 2004). The pollution potential of the runoff depends on factors such as tree species, the amount of water that is in contact with the wood, whether irrigation water is collected and recirculated, if the runoff is treated to reduce the concentration of pollutants, and the size and condition of the receiving watercourse. If the runoff is released to a small watercourse, it will have a considerable negative effect on the ecosystem simply due to the resulting comparatively large proportion of polluted water (Hedmark and Scholz 2008).

2.5.2.2 Characteristics of Runoff

Because of dissimilar conditions at different log yards, scientific studies have come to different conclusions concerning the pollution potential of runoff from wood handling sites. There are considerable differences in the characteristics of the runoff. These differences exist for various reasons; *e.g.*, the species of tree stored on the log yard, the proportion of the runoff that has come in contact with stored wood, and the dilution or pretreatment of the runoff before sampling. Different species of trees contain varying concentrations and types of soluble compounds, and the ease with which the resultant extractives are leached from the wood also greatly influences the runoff concentration and the corresponding toxicity (Zenaitis *et al.* 2002). It is also reasonable to assume that the duration of time for which the wood has been stored before the samples are taken influences the concentration of pollutants in the runoff because of degradation processes in the wood (Feist *et al.* 1971).

Despite the differences between studies, some general conclusions can be drawn from the characteristics of various runoffs. It appears that the main problem connected with pollution from log yard runoff is the high amount of organic matter that results in increased oxygen demand when it degrades in the receiving watercourse (Jonsson *et al.* 2006). Some of the organic compounds may also be toxic to aquatic plants and animals. In addition, the increased concentrations of phosphorus in log yard runoff might lead to considerable loads in the large runoff volumes produced from a sprinkled log yard every year (Hedmark and Scholz 2008).

The amount of organic matter in runoff is generally relatively high if compared to concentrations in the receiving watercourses (*e.g.*, www.environment-agency.gov.uk). The published studies show that water-soluble compounds present in bark and wood are released and appear in high concentrations in the corresponding runoff.

Some researchers have gone into detail to find out which organic substances are present in runoff (Borgå *et al.* 1996a; DeHoop *et al.* 1998; Field *et al.* 1988; Tao *et al.* 2005; Taylor *et al.* 1996; Wang *et al.* 1999). The compounds that are generally found are phenols, resin acids, tannins and lignins, and volatile fatty acids. These compounds may be toxic to aquatic plants and animals. Of the organics extracted from softwood, tannins, lignins, phenols, tropolones, and resin acids are of greatest concern because they are seen to be contributors to runoff toxicity (Samis *et al.* 1999). Two studies (Bailey *et al.* 1999; Doig *et al.* 2006) also found elevated concentrations of metal (zinc and aluminum, and zinc, respectively). In the study by Bailey *et al.* (1999), the high concentrations of zinc were thought to originate from buildings in the studied sawmill area.

The amount of particles in the runoff can also be relatively high. A study in British Columbia, Canada, focused on the characterization of particles in fresh and primary treated log sort yard runoff and found that the suspended and colloidal particles in the runoff were largely organic in nature (Doig *et al.* 2006).

Concentrations of phosphorus can be sufficiently high as to cause eutrophication in the receiving watercourses (Scholz *et al.* 2007). Another common characteristic of runoff is that the concentration of nitrogen is generally relatively low if compared to corresponding concentrations in natural watercourses (Scholz 2006a). The pH has been shown to be very low in some studies (*Thuja plicata* (red cedar), *Chamaecyparis nootkatensis* (yellow cedar), and *Populus tremuloides* Michx. (trembling aspen)), and neutral in other studies (*Picea abies* (L.) Karst (Norway spruce), and *Pinus sylvestris* L. (Scots pine)). According to studies where different tree species have been directly compared with each other, there are differences in the runoff generated. For example, emissions from storage of *Picea abies* are generally greater than those from *Pinus sylvestris* (Borgå *et al.* 1996b). In leaching experiments reported by Pease (1974), the relative COD leaching rates of four species of trees were, in descending order, *T. plicata*, *C. nootkatensis*, *Tsuga heterophylla* (hemlock), and *Picea sitchensis* (Sitka spruce).

An interesting finding by Borgå *et al.* (1996b) is that the release of pollutants from the stored wood might be affected by the chemistry of the water that comes in contact with the wood, in this case, sprinkling water. The nutrient content of the water used for sprinkling had a significant effect on the pollutants emitted from stacks of *Picea abies* and *Pinus sylvestris*. This was explained by the dependency of the growth of bacterial biomass in the log piles. Rapid growth of bacterial biomass caused by high concentrations and loads of nutrients in the sprinkling water reduced the initially high environmental loadings from the wood.

Many geoclimatic, operational, and physical factors contribute to the volume and characteristics of runoff, and a management tool to predict relative environmental risks from different sites would be of considerable value. In a study by Orban *et al.* (2002), the authors attempted to develop such a tool. A survey was devised and distributed to log yard operators in British Columbia, Canada. The survey provided information on site characteristics, volumes and types of wood processed, operational practices, incidence of runoff, runoff treatment practices, and the ultimate receiving environment. Multidiscriminant analysis was used to determine which factors were correlated to environmental risk posed by runoff. In order of importance, volume of wood stored onsite (largest contribution), frequency of runoff events, and color intensity of runoff (smallest contribution) were factors that significantly contributed to risk and were correlated positively. The methods used in this study could be applied as tools to evaluate the need for remediation at different log yards.

2.5.2.3 Effects of Runoff on Receiving Watercourses

Pollutants in runoff influence the receiving watercourse in different ways, largely depending on the geographical location, size, and chemical and biological characteristics of the watercourse. Organic material can cause oxygen depletion when it is biologically and chemically degraded, organic compounds and metals can have a toxic effect on plants and animals, nutrients in the runoff can lead to eutrophication, and particles in the runoff can have a negative impact on plant and animal life due to clouding of the water and sedimentation on substrates (Scholz 2006a; Hedmark and Scholz 2008).

A study of invertebrates in the Sapele stretch of the Benin River (Niger delta, Nigeria) found that sensitive species of genera such as *Ephemeroptera* (mayfly) or *Plecoptera* (stonefly) were absent from the location where wood waste discharge was emitted into the river (Arimoro and Osakwe 2006). The wood waste discharge altered the water chemistry and stimulated the abundance of less-sensitive macroinvertebrate species.

Dumping and rafting logs in water also leads to a potentially serious environmental impact. A study on water quality and benthic organisms at 16 marine sites in southeast Alaska (USA) showed a layer of bark covering the bottoms at all log dumping sites, and this layer was associated with a high biological oxygen demand (Pease 1974). The benthic epifauna was reduced in abundance only at the oldest active log dump studied, but the bottom fauna was reduced in all the bark-covered areas. Organic compounds leached rapidly from all species of log studied but were precipitated by the seawater. Significant effects on water quality were found at only two of the sites.

Some studies also report toxic effects, which seem to be closely related to the tree species investigated. In leaching experiments for *T. plicata*, *C. nootkatensis*, *T. heterophylla*, and *P. sitchensis* (Pease 1974), *P. sitchensis* and *T. plicata* were found to be the most toxic to *Oncorhynchus gorbuscha* (pink salmon) fry in freshwater, and *C. nootkatensis* was found to be the most toxic in seawater. All wood species were found to be more toxic in freshwater than in seawater.

Populus tremuloides logs in Canada have been reported to produce a dark, watery, and acutely toxic leachate (Taylor *et al.* 1996). Median acutely toxic concentrations of leachate were consistently between 1 and 2% of its full strength for *Oncorhynchus mykiss* (rainbow trout) and *Daphnia magna*. Inhibition of bacterial metabolism (Microtox) began at concentrations below 0.3%. The leachate was less toxic to plant life but inhibited growth at concentrations between 12 and 16%. Oxygen depletion, low pH, and phenolic compounds contribute to the toxicity of *P. tremuloides* leachate, but much of the toxic effect must be attributed to other unidentified constituents (Taylor *et al.* 1996).

Another Canadian study (Bailey *et al.* 1999) found that runoff from nine sawmills in British Columbia was toxic to juvenile *O. mykiss*, mainly because of the content of zinc, tannins, and lignins. A third study (Taylor and Carmichael 2003) comprising tests on *O. mykiss*, *D. magna*, and luminescent bacteria (Microtox) showed that leachate from *P. tremuloides* varied from weakly toxic (median lethal concentration >10%) to very toxic (median lethal concentration <1%) levels.

A study on *T. plicata* and *C. nootkatensis* wood waste including trimmings, off-specification wood chips, shredded bark and roots, and sawdust (Tao *et al.* 2005) showed that the runoff was very toxic to aquatic life with a 96-h median lethal concentration of 0.74% leachate. According to Tao *et al.* (2007), the acute toxicity of wood leachate is usually attributed to tannins, lignins, tropolone, terpene, lignans, and low pH.

A study on log yard runoff from *Picea abies* and *Pinus sylvestris* found distillable phenols and diterpene resin acids that are potentially toxic to aquatic life (Borgå *et al.* 1996a). However, in a Swedish study (Arvidsson 2006), undiluted runoff from a sprinkled log yard at a sawmill storing the same two species was found not to have any negative effect on the reproduction of *Daphnia sp.*.

It is difficult to directly compare the environmental impacts of pulp and paper mill effluents with the corresponding impacts from sawmills, log yards, and wood waste handling sites. The sizes of the sites and the corresponding scales of operation are generally greater at pulp and paper mills compared to other wood handling sites, and the use of chemicals in the processes is more intense. The impacts of the effluents on the local watercourses, however, have a lot in common because of the comparably high content of organic compounds from the wood handled at the sites. With this in mind, a few representative studies on the effects of the organic matter in effluents from pulp and paper mills on local watercourses have been included in this section.

Karrasch *et al.* (2006) investigated the impact of pulp and paper mill effluents on the water quality, microplankton system, and microbial self-purification capacity (*i.e.*, degradation of polymeric organic compounds via extracellular enzymes) of the Biobio River, Chile. They found that the impact of the effluents on the water quality was indicated by raised conductivity, increased concentrations of nitrate, nitrite, and SRP, the appearance of tannins and lignins, and a steady accumulation of inorganic and organic suspended matter along the river. Very low and declining concentrations of chlorophyll *a* and heterotrophic flagellate densities were reported. The pulp and paper mill effluents resulted in high bacterial abundances and biomass concentrations in the river water, showing that bacteria adapted to the effluent made effective use of the available nutrients and that the grazing pressure from heterotrophic flagellates was reduced (Hedmark and Scholz 2008).

Furthermore, Karrasch *et al.* (2006) showed that a pulp mill contributed to the self-purification of an affected river stretch. However, the elevated degradation capacity was not sufficient to compensate for the high load of organic material in the paper plant effluent, which, together with its toxic effects, significantly interfered with the ecological status of the Biobio River.

2.5.3 Treatment Methods

2.5.3.1 Overview of Applied Treatment Technologies and Methods

Several different technologies such as biological systems (*e.g.*, aeration lagoon, activated sludge process, and constructed wetland) and physical and chemical systems (*e.g.*, aeration, carbon adsorption, coagulation, ion exchange, neutralization, precipitation, reverse osmosis, chemical oxidation using ozone, calcium hypochlorite, hydrogen peroxide, and potassium permanganate, and chelation) have been proposed by Orban *et al.* (2002) and Samis *et al.* (1999) for the treatment of runoff from wood handling sites. Given the high load of organic compounds in the runoff, any method used to effectively treat this wastewater has to allow for the degradation of these compounds.

Different types of soil–plant systems and wetlands (Scholz and Lee 2005; Scholz *et al.* 2007) might also be used to treat log yard runoff with high concentrations of phosphorus. When the elevated level of phosphorus in the runoff is a problem, an efficient purification method must reduce the concentration of phosphorus in the water before it reaches the receiving watercourse. The methods that have been used in laboratories or field-scale trials have focused mainly on the degradation of the organic matter in the runoff and the accompanying reduction of toxicity.

2.5.3.2 Soil Infiltration

The reduction of pollutants that is achieved by soil infiltration of runoff is due to physical, chemical, and biological processes taking place in the soil–plant water ecosystems (Wang *et al.* 1999). A soil infiltration system example is shown in Figure 2.24. The degree of degradation, assimilation, sorption, exchange, and neutralization during the infiltration also depends on the velocity of the solution through the soil. Biological degradation, mineralization, and adsorption are the

Figure 2.24 Experimental soil infiltration system treating log yard runoff near Heby, Sweden

major mechanisms contributing to the reduction of organic substances in the soil column. If contaminants are predominantly retained through adsorption alone, then a breakthrough of high concentrations of organics in the leachate can occur when the soil adsorption capacities are exceeded (Wang *et al.* 1999).

An early study by Peek and Liese (1974) in Lower Saxony (Germany) concluded that the purifying capacity of the soil in the area where windthrown timber was stored and sprinkled for a year was sufficient to prevent a decline of water quality in an adjoining well or stream. More recent studies have shown that infiltration of log yard runoff in unlined wetlands planted with different plant species have been successfully used for a sawmill in central Sweden (Hedmark and Jonsson 2008), even at irrigation intensities of up to 66 mm/d. The purification efficiency of the infiltration wetland was continuously high for total organic carbon, total phosphorus, and phenols, and the purification capacity was clearly maintained after 4 years of irrigation. During the fourth year of operation, however, the concentrations of total phosphorus in the groundwater below the irrigated area increased significantly compared to previous years. This may indicate that the soil is close to the saturation point for total phosphorus (Casson *et al.* 2006).

Other reports by Jonsson *et al.* (2004, 2006) based on soil infiltration experiments in field-scale studies utilizing lysimeters have shown good purification results. There was no difference in purification efficiency for total organic carbon, distillable phenols, total phosphorus, and total nitrogen when comparing treatments by *Elytrigia repens* (L.) (couch grass), *Salix sp.* (willow), and *Alnus glutinosa* (L.) Gärtner (alder) with each other. No clear differences were found between lysimeters containing sand or clay. Lysimeters with high levels of irrigation showed greater retention than those with low levels.

2.5.3.3 Wetland Treatment

Constructed wetlands are recognized as low cost and 'minimal' maintenance systems that could lower the impact of wastewater or storm water discharges on natural water bodies (Scholz 2006a; Scholz and Lee 2005). Wetlands, ponds, lagoons, and trenches (Figure 2.25) are sometimes used at sawmills and log yards to enhance treatment of runoff including sedimentation of particles. For example, a vegetated trench is used for treatment of log yard runoff from a sawmill in Boxholm, Sweden. This wetland performs well in terms of organic matter removal.

In a study by Doig *et al.* (2006), the composition of the organic matter in the runoff from a log sort yard was examined before and after treatment in a lagoon. The fresh runoff had chemical oxygen demand concentrations ranging between 346 and 3690 mg/l. In a rainfall-generated sample, particulates (*i.e.*, particles greater than approx. 1.5 μm) contributed up to 52% of the total COD, and colloids (particles between 0.02 μm and approx. 1.5 μm) were associated with only 39% of the corresponding COD. Following primary treatment in a lagoon, the organic particles found were mostly of a colloidal nature. Primary treatment experiments showed that between 27 and 54% of the COD could be removed by settling, depending on the initial concentration. An additional 33% of COD was removed, probably due to biological degradation during settling. The COD of the ultimately treated effluents remained relatively high (378 to 533 mg/l), indicating that not all suspended material could be removed through settling and biodegradation alone, and that the application of further treatment methods would be required.

The use of vegetated wetlands usually leads to better purification results than lagoons or ponds (Scholz 2006a). There are various reasons for the better efficiency. For example, macrophytes act as mechanical filters and provide substrate for microorganisms, which take part in degradation and absorption processes of pollutants. Furthermore, aquatic plants aerate water and the root zone, and reduce the flow speed of the water, giving particles more time to settle. Sediment adsorption is likely to affect the treatment performance in surface-flow constructed wetlands predominantly during the initial and transitional operating periods. However, the major mechanism for the removal of organic carbon from wood waste leachate during long-term operation is likely to be biological degradation (Tao *et al.* 2007).

Tao *et al.* (2006) studied the treatment performances of four vegetated surface-flow mesocosm wetlands fed with different dilutions of wood waste leachate over a period of 12 weeks. The wood waste leachate was diluted before discharge to constructed wetlands to prevent potentially adverse impacts of the raw leachate on heterotrophic bacteria and aquatic plants. During a subsequent period of 13 weeks, the effluent of a vegetated wetland fed with the raw leachate was further treated in

Figure 2.25 Wetland trench treating log yard runoff near Boxholm, Sweden

both a vegetated wetland and an open wetland. The highest reduction rates for COD as well as tannins and lignins were achieved in the wetland fed with the raw leachate. The most diluted (up to six times) wood waste leachate yielded the highest reduction efficiencies. Up to 47 mg/l volatile fatty acids in the influent were depleted through wetlands with a hydraulic retention time of 13 d. In this study, the vegetation made an insignificant difference in terms of the treatment performance for wood waste leachate. The COD as well as tannins and lignins were further removed through the wetlands operating in series, though at lower reduction rates. The performance of wetlands for the treatment of wood waste leachate was partly regulated by the dissolved oxygen concentration and the availability of bacterial substrates.

In another field experiment, Tao *et al.* (2006) studied how different hydraulic retention times affected the purification efficiency of four surface-flow mesocosm wetlands. The wetlands were fed with diluted wood waste leachate, which was acidic, of very high oxygen demand, and toxic. Mass reduction efficiencies of COD, tannins, and lignins increased significantly with hydraulic retention time when 10% leachate was diluted with tap water. When a more recalcitrant influent was fed into the wetlands, there was a slight increase of reduction efficiency with increasing hydraulic retention time. Precipitation and evapotranspiration had profound impacts on the overall performance and its variability.

Another study on the effectiveness of constructed wetlands for the treatment of wood waste leachate was conducted by Masbough *et al.* (2005) on a wood waste storage site, adjacent to the Lower Fraser River (near Mission, British Columbia, Canada). The leachate was characterized by high oxygen demand, tannins and lignins, and volatile fatty acid concentrations, but by low pH and nutrient values. Diluted leachate passed through six pilot-scale wetland cells, four planted with *Typha latifolia* L. (reedmace) and two unplanted controls, with a mean depth of 0.4 m and a hydraulic retention time of 7 d. Nutrient addition and pH adjustments were made to improve contaminant removal. Reductions in contaminants were consistently achieved, with mean removals for BOD, COD, volatile fatty acid, and tannins and lignins of 60, 50, 69, and 42%, respectively. Furthermore, aging of the constructed wetland system increased the treatment performance.

2.5.3.4 Other Treatment Methods

Where it is not possible to construct wetlands or use infiltration as a treatment method, for example due to limited land availability, more high-tech methods can be used to reduce the impact of contaminated runoff from wood handling sites. There are only a limited number of studies so far concerning runoff from log yards and sawmills. More studies have been conducted on wastewater from pulp and paper mills, and those are discussed in Section 2.5.3.5.

In a study by Woodhouse and Duff (2004) on runoff from a sawmill on Vancouver Island (British Columbia, Canada), five runoff samples were treated in a laboratory-scale attached microbial growth reactor. Six samples were acutely toxic

(EC50 < 100%) based on the Microtox assay. The samples were effectively treated in the attached microbial growth reactor. Treatment for 24 h at 34°C resulted in substantial reductions of BOD, COD, and tannins and lignins. Near complete removal of acute toxicity and color were also observed. Twenty-four-hour treatment at lower temperatures (24°C and 5°C) also substantially reduced the concentrations of organic matter and the toxicity.

In another Canadian study, samples of log yard runoff were obtained from two British Columbia coastal sawmills (Zenaitis and Duff 2002). Centrifuged samples were treated with ozone doses up to approx. 0.5 mg ozone/mg COD in a lab-scale reactor. Ozonation was found to significantly reduce toxicity, tannin and lignin, and dehydroabietic acid concentrations. However, there were only moderate reductions in COD and BOD. At slightly acidic to neutral conditions, pH had no effect on the COD. Tannin and lignin, and toxicity removal were slightly improved in neutral solutions compared to acidic ones, while dehydroabietic acid removal significantly improved.

In another study on Vancouver Island (British Columbia, Canada), ozone treatment of log yard runoff was combined with biological treatment (Zenaitis et al. 2002). Batch biological treatment of log yard runoff reduced the BOD, COD, and tannin and lignin concentrations considerably. Acute toxicity decreased from an initial EC_{50} of 1.83% to a value of 50.4% after 48 h of treatment.

The efficiency of ozone as a pre- and postbiological treatment stage was also assessed by Zenaitis et al. (2002). During ozone pretreatment, the combined tannin and lignin concentration and the acute toxicity were rapidly reduced. Preozonation had a minor reduction effect on the BOD and COD. However, a larger fraction of residual COD was non-biodegradable after ozonation.

Moreover, biologically treated effluent was subjected to ozonation to determine whether further improvements in effluent quality could be achieved. Reductions in COD and in tannin and lignin concentrations were observed during ozonation. However, no further improvement in toxicity was observed. Ozonation increased the BOD due to chemical conversion (Zenaitis et al. 2002).

2.5.3.5 Treatments Used for Organic Matter in Pulp and Paper Mill Wastewater

In this section, the focus will be on techniques that are applicable for the treatment of organic matter contained within log yard runoff and wood waste leachate as well as within effluents from pulp and paper mills. The effluent characteristics of pulp and paper mills can differ considerably from those of the runoff from wood handling sites because of the chemicals used in the production processes. However, a relatively large proportion of organic material in the wastewater originates from the wood handled at the sites (Wang et al. 1999).

The wastewaters from mechanical pulping and secondary fiber pulping and the condensates from chemical pulping are non-toxic to methanogenic degradation and contain easily degradable organic compounds. Anaerobic digestion is therefore an attractive treatment for these effluents (Rintala and Puhakka 1994). This technology, often in combination with aeration units, is now well established in the industry.

Munoz *et al.* (2006) applied different advanced oxidation processes to remove the organic carbon content of a paper mill effluent originating from a pulp bleaching process. The oxidation processes comprised titanium-dioxide-mediated heterogeneous photocatalysis, titanium-dioxide-mediated heterogeneous photocatalysis assisted with hydrogen peroxide, titanium-dioxide-mediated heterogeneous photocatalysis coupled with Fenton, photo-Fenton, ozonation, and ozonation with ultraviolet-A light irradiation. The results showed that all selected oxidation processes considerably decreased the dissolved organic carbon content.

Moreover, a life cycle assessment indicated that heterogeneous photocatalysis coupled with the Fenton's reagent had the lowest environmental impact and a moderate to high dissolved organic carbon removal rate. On the other hand, heterogeneous photocatalysis appeared to be the worst oxidation process both in terms of dissolved organic carbon abatement rate and corresponding environmental impact. The environmental impact of the studied advanced oxidation processes is attributable to the high electrical energy consumption necessary to run a UV-A lamp or to produce ozone (Munoz *et al.* 2006).

Nair *et al.* (2007) studied the effect of phenol degrading *Alcaligenes* spp. on paper factory effluent. Three techniques were applied: use of free cells, immobilized cells, and a continuously operated packed bed reactor. An almost complete reduction of phenol was found for all techniques. When comparing them with each other, the reduction capacity for COD was the best with the continuously operated packed bed reactor, even at a much shorter treatment time. After the complete removal of phenol, the strain could still further enhance the reduction of the COD, which clearly indicated that this strain could also oxidize organic matter other than phenol in the paper factory effluent.

Uğurlu *et al.* (2006) studied the effectiveness of electrochemical treatment for the high concentration of organic pollutants in paper mill effluents. They used a working electrode cell consisting of a graphite electrode and powder-based activated carbon. Electrolysis time, voltage, initial pH, activated carbon, sodium chloride concentration, and air flow were selected as parameters. The removal efficiency significantly depended on the applied cell voltage, air flow, time, salt concentration, and pH. COD and lignin and phenol removal efficiencies from the paper mill effluent were just below 90% after electrolysis.

Ozonation has been evaluated for the treatment of both log yard runoff and pulp mill effluents. In a study by El-Din *et al.* (2006b), ozonation of resin and fatty acids, which were found in pulp mill effluents, was assessed using rapid-scan stopped-flow spectrophotometry. The resin and fatty acid oxidation (*i.e.*, degradation) efficiency increased with an increase of ozone and temperature. Microtox bioassay tests showed that there was an increase in toxicity for resin acid as a result of ozonation. The toxicity of fatty acid samples, however, decreased as a result of ozonation.

In a further study by El-Din *et al.* (2006a), ozonation of abietic and linoleic acids (resin and fatty acids) was assessed at various pH levels and low temperatures

(0 to 10°C) using a rapid-scan stopped-flow spectrophotometer. Degradation of abietic and linoleic acids could be enhanced by manipulating reaction conditions including ozone dose, temperature, and pH. Ozonation of abietic acid led to a 140% increase in Microtox toxicity, whereas it was reduced to half in the case of linoleic acid.

Wetlands have also been investigated for the treatment of wastewater from pulp and paper mills. The capacity of a pilot project sub-surface batch flow constructed wetland in the tropics to remove phenol from pretreated pulp and paper mill wastewater was studied under varying hydraulic retention times (Abira *et al.* 2005). Initial results based on a 15-month study indicated that mean removal efficiencies for phenol were variable but reached 60 and 77% at 5-d and 3-d hydraulic attention times, respectively. It was thought that the longer retention time might have caused oxygen and nutrient deficiencies, which may have reduced the removal performance. Although phenol was sometimes not detectable in the wetland outflow, mean values over the experimental period did not meet national guidelines.

Irrigation and infiltration of pulp mill effluents have also been investigated. For example, large barrel lysimeters containing intact soil cores were irrigated with thermomechanical pulp effluent in New Zealand for 16 months (Wang *et al.* 1999). The thermomechanical pulp mill used *Pinus radiata* D. Don (radiata pine) as its main wood source. The primary treated effluent was applied on the lysimeters at a loading rate of 30 mm/week. Analysis of the effluent showed low concentrations of nutrients and inorganic contaminants, but high concentrations of wood-derived organic contaminants such as total organic carbon, COD, BOD, and wood extractives. Removal rates through the soil core were greater than 90% for total organic carbon, COD, BOD, turbidity, resin acids, phytosterols, and phenols. No excessive nutrient and heavy metal leaching was observed during the experimental period.

2.5.4   Discussion, Conclusions, and Further Research

2.5.4.1   Discussion Concerning the Cost-effectiveness of the Treatment Methods

The literature review presented here has highlighted that runoff from wood handling sites can cause negative environmental effects in receiving watercourses. However, it is common in industry that no treatment of wastewater is done prior to its release. As knowledge regarding the content and the environmental impact of the runoff increases, environmental authorities are demanding more frequently that the pollution loads from the sites be reduced. Low-cost biotechnologies such as constructed wetlands for the treatment of runoff have gained considerably in popularity (Scholz 2006a).

The review presented here indicates that for sawmills and log yards, it is important to select a treatment method with reasonable costs for installation and maintenance. Treatment with constructed wetlands or infiltration in soil and plant systems has been shown to be effective for reducing the concentrations of organic matter and phosphorus and for reducing the toxicity of the runoff from wood handling sites. These methods do not require high technical maintenance, and the amount of labor necessary to keep them operating is low. The cost of construction, however, depends on the soil properties and the cost of land at each site.

Constructed wetlands are less costly than conventional techniques used for the treatment of wastewater and are a viable option for wood handling operators. At sites with a shortage of land for constructed wetlands or soil infiltration, it is likely to be necessary to invest in treatment methods that are more costly, energy demanding, or high-maintenance such as ozonation.

The processes used to treat the organic matter in pulp and paper mill effluents are also applicable to treat runoff from wood storage sites. However, the cost of the treatment may be an important limiting factor. Based on the author's experiences, sawmills and other wood handling sites usually have a smaller turnover than pulp and paper mills and therefore do not have the same means to invest in high-tech water treatment equipment.

Based on the author's professional experience, ozonation is generally associated with very good purification efficiencies for organic matter and can also reduce the accompanying toxicity of these compounds. Ozonation of some compounds such as resin acid can lead to increased toxicity. However, this depends on the formation of degradation products of higher toxicity than the original compounds.

The degree of purification that needs to be achieved largely depends on the condition of the receiving watercourse and demands from the environmental authorities and the public. It follows that different treatment methods alone or in combination with each other (treatment train) might be fit to a given purpose (Scholz 2006a).

Constructed wetland biotechnology could be enhanced by taking advantage of recent advances in microbiology. For example, Doradoa et al. (2000) indicated the potential of wood pretreatment with selected sapstain fungi for decreasing effluent toxicity in pulping (Doradoa et al. 2000). The combination of this pretreatment method with constructed wetlands is likely to lead to a reduction in required wetland area, making this biotechnology more attractive for areas where land is costly.

2.5.4.2 Summary of Conclusions

This section addresses the timely and urgent need of the wood and sawmill industries to assess various soft technologies including wetland systems to treat and attenuate log yard runoff. Moreover, the pollution potential of runoff from wood handling sites has been assessed. Organic matter and phosphate are the key contaminants.

The reasons for pollution generation at wood handling sites have been highlighted, and the runoff and its effect on the receiving watercourses have been assessed. Soft treatment biotechnologies including wetland and soil infiltration have been characterized and assessed. Various wetland systems including a novel combination of a size-optimized wetland with a runoff conveyance channel (case study in Boxholm, Sweden) have been suggested for future application. The proposed system reduces phosphates and organic matter to concentrations below common wastewater treatment guidelines. The key advantage of most relevant wetland and soil infiltration systems is that they require little energy and low maintenance.

Ozonation has been shown to often effectively decrease acute toxicity, but to have little effect on the COD and BOD. Ozonation of some organic compounds (*e.g.*, resin acid,) can also lead to increased toxicity. Therefore, this high energy-consuming advanced oxidation technology is often unsuitable for the treatment of log yard runoff.

Finally, various treatment methods used for the removal of organic matter in pulp and paper mill wastewater have been assessed. Based on this review, wetland systems similar to those treating log yard runoff are recommended for the industry.

2.5.4.3 Further Recommended Research

The environmental impact from each separate wood handling site is relatively small, but cumulatively, the overall environmental impact from the storage and handling of wood is likely to be of high significance in terms of the pollution load. This hypothesis needs to be tested in future research studies.

It is also common that irrigation water for a log yard is abstracted from a watercourse close to the yard, and in small watercourses overabstraction can lead to drying out or a significant reduction of the water level, which may have negative effects on both the biology and the chemistry of the watercourse. However, no research on this aspect of the wood handling industry's environmental impact can be found in the public domain.

This section supports the reader in choosing a sustainable and cost-effective treatment technique for runoff from wood handling sites. However, no ultimate method that gives sufficient long-term removal of phosphorus was found. Further studies on sustainable treatment methods that can effectively remove phosphorus from runoff are therefore strongly recommended.

Further research should be undertaken on the 'real' whole life costs and tangible benefits of wetland treatment systems compared with traditional methods such as the activated sludge process and percolation filtration. Moreover, the modeling of wetland performance for highly dynamic (*i.e.*, seasonal operation and fluctuating stock capacities) systems such as log yards remains a challenge.

2.6 Wetlands for Treating Hydrocarbons

2.6.1 Introduction

2.6.1.1 Constructed Treatment Wetlands

Wetlands are complex and integrated systems in which water, animals, plants, microorganisms, and the environment interact to improve the water quality (Guirguis 2004). Man has started to mimic nature by building wetlands to treat a variety of waters, wastewaters, storm waters, gully pot liquor, acid-mine drainage waters, landfill leachate, irrigation waters, agricultural wastewater, runoff waters, industrial wastewater, and produced waters (Moshiri 2000; Rew and Mulamoottil 1999; Scholz 2004; Vrhovsek et al. 1996; Yang et al. 2001). However, very little is known about the processes involved and the use of constructed treatment wetlands in the removal of petroleum hydrocarbons from processed water. This has limited the effective application of this technology in the oil and gas industries.

Vertical-flow and horizontal-flow constructed reed beds based on soil, sand, or gravel are used frequently to treat domestic and industrial wastewater (Kadlec and Knight 1996). Constructed wetlands for wastewater treatment have several advantages such as low operational costs and habitat enhancement over conventional treatment methods (Cooper et al. 1996; Moshiri 2000). The wetland treatment performance depends on the interactions between many different physical and biochemical components. It follows that some functions such as the role of aggregates (also called substrate or wetland media) within constructed wetlands are not completely understood (Scholz 2006a).

Vertical-flow wetland systems have a structural makeup of several layers of aggregates. Water is forced to flow perpendicular to the length of the wetland. Wetlands consist of different sizes of gravel and sand and are often planted with macrophytes, which provide oxygen to the rhizosphere, thereby creating an aerobic environment (Scholz 2006a). This in turn supports microbial communities, which can either directly biodegrade organics or catalyze chemical reactions and subsequently support biotransformation processes. Bacteria capable of degrading volatile organics such as benzene, toluene, ethylbenzene, and p-xylene have been found in the rhizosphere (Hiegel 2004; Sugai et al. 1997). Numerous benzene-degrading aerobic microorganisms have been identified; the most notable are the *Pseudomonas* spp., which may account for up to 87% of the gasoline-degrading microorganisms in contaminated aquifers (Ridgeway et al. 1990). Petroleum wastes have been documented to degrade in natural wetland environments (Wallace and Knight 2006; Wemple and Hendricks 2000). Benzene is biodegradable, particularly in the presence of oxygen (Wemple and Hendricks 2000). Benzene degradation has also been demonstrated in the presence of nitrate–nitrogen (Burland and Edwards 1999). Many studies have shown that microbial degradation of petroleum hydrocarbons in the environment is strongly influenced by physical and chemical factors such as temperature, oxygen, nutrients, salinity, pressure, water activity, and pH and the chemical composition, physical state, and concentration of the contaminant, as well as by biological factors such as the composition and adaptability of the microbial population (Burland and Edwards 1999).

Kadlec (2001) states that aeration is an important component of sub-surface-flow wetland design because an active aeration system enhances both volatilization and aerobic degradation of hydrocarbons. The transfer of oxygen to contaminated aquifers to stimulate aerobic degradation is a common bioremediation practice (Lovley and Lloyd 2000). The biodegradability of the most water-soluble components of gasoline such as benzene (Paje et al 1997; Solano-Serena et al. 1999), toluene (Leahy and Olsen 1997), ethylbenzene (Di Lecce et al. 1997), and xylene (Di Lecce et al. 1997) compounds has been clearly established.

Because biodegradation and evaporation processes compete in removing petroleum hydrocarbons, biodegradative losses cannot be differentiated clearly from volatility losses (Zhou and Crawford 1995). Mitsch and Gosselink (1993) revealed that freshwater wetlands are typically considered to be nutrient-limited due to the heavy demand for nutrients by the plants, and they could also be nutrient traps, as a substantial amount of nutrients may be trapped in biomass. Hence the addition of nutrients is necessary to enhance the biodegradation of oil pollutants (Choi et al. 2000; Kim et al. 2005). However, studies in the past (Chaillan et al. 2006) have shown that excessive nutrient concentrations can inhibit biodegradation activity, and several authors have reported the negative effect of high nitrogen, phosphorus, and potassium levels on the biodegradation of hydrocarbons (Chaineau et al. 2005; Oudot et al. 1998), and particularly on the aromatics (Carmichael and Pfaender 1997).

2.6.1.2 Benzene Removal

There is a gap in knowledge concerning the optimization of those process conditions that would foster a more efficient application of the constructed wetland technology particularly for the removal of hydrocarbons under different climatic and other variable environmental conditions. However, the number of research projects concerning constructed wetlands for hydrocarbon treatment has considerably increased (Baris et al. 2001; Eke and Scholz 2006; Myers and Jackson 2001; Omari 2003; Xia 2003).

Since 1995, journal articles and symposia proceedings indicate the petroleum industry's interest in using constructed wetlands to manage process wastewater and storm water at a variety of installations including refineries, oil and gas wells, and pumping stations (Knight 1999). The area of emphasis in this research is the use of constructed wetlands for the treatment of dissolved petroleum hydrocarbon compounds, hence the need to have an understanding of what constitutes petroleum and its degradation products. There are four basic petroleum compounds (Harayama et al. 2004):

- saturated hydrocarbons, the primary component;
- aromatic hydrocarbons (at least one aromatic ring within the molecule);

- resins, which are mostly unknown; and
- asphaltenes.

These compounds are relatively large molecules and tend to posses small amounts of nitrogen, sulfur, or oxygen (Harayama *et al.* 2004). Untreated petroleum industry wastewaters contain phenolics (Knight 1999), sulfides, and various trace metals (Knight 1999). Hydrocarbons are highly soluble and neurotoxic and cause cancer (Hiegel 2004). Monoaromatic for benzene, toluene, ethylbenzene, and xylenes (BTEX), hydrocarbons (*i.e.*, benzene, toluene, ethylbenzene, and xylene) are commonly found in gasoline and are highly volatile substances (Coates *et al.* 2002). Due to their relatively high solubility and toxicity, they represent a significant health risk in contaminated environments. Of all the BTEX compounds, benzene poses the gravest concern because it is the most toxic and a well-known human carcinogen. The benzene ring (Figure 2.26) is a chemical structure that is common in nature. Moreover, the thermodynamic stability of the benzene ring increases its persistence in the environment; therefore, many aromatic compounds are major environmental pollutants (Dagley 1986; Díaz Eduardo 2004).

Their major industrial source is petroleum and natural gas, formed geochemically from biomass under high pressure and temperature (Heider *et al.* 1998). Aromatic hydrocarbons are one of the most abundant class of organic compounds and constituents of petroleum and its refined products. Monocyclic aromatic hydrocarbons are of major concern, because of their toxicity, high solubility, and ability to migrate within groundwater. These BTEX compounds are of primary discharge concern for the water quality of receiving waters (Caswell *et al.* 1992). The BTEX fraction of total volatile hydrocarbons is primarily responsible for most of the total toxicity in gasoline-contaminated groundwater. Hence, an attempt to reduce toxicity requires targeting these compounds for destruction. Many components of hydrocarbon mixtures are toxic and relatively soluble in water. In natural gas, benzene concentrations typically range from about 0 to 1,000 mg/l; in crude oils from virtually zero to 10,000 mg/l (Janks and Cadena 1991). Benzene has relatively high water solubility (1,780 mg/l). Water contamination by oil exploration and production operations, tank farms, underground storage tank leakage, and refineries has become a concern to the oil and gas industry.

Hydrocarbon degradation is less dependent on the actual reactions taking place than it is on the processes occurring in the surrounding ecosystem (Sugai 1997). Aerobic biodegradation and volatilization constitute a coupled pathway that contributes significantly to the natural attenuation of hydrocarbons (Lahvis *et al.* 1999).

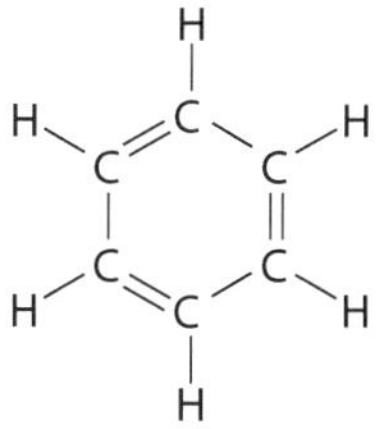

Figure 2.26 Chemical structure of benzene

Achieving high treatment performances within a short time is critical, so the design can be amended to allow manipulations of environmental conditions to enhance dissolved hydrocarbon treatment. Environmental conditions to be taken into consideration include dissolved oxygen (DO), pH, temperature, and nutrient requirements (*i.e.*, nitrogen and phosphorus) of the wetland plants and microbes. Associated contamination is therefore a major environmental problem due to the manufacture, transportation, and distribution of petroleum (Atlas and Cerniglia 1995).

2.6.1.3 Novelty, Aim, and Objectives

The research uses data gathered from experimental small-scale wetlands to assess the efficiency of benzene removal in each wetland and to compare different operational conditions. The research covers the assessment of environmental, physical, and microbial processes. This enhances operational knowledge and understanding of treatment wetlands and provides data that could be used to design full-scale wetland systems for efficient hydrocarbon treatment, and to model biodegradation and operational processes. Improved system control should include knowledge and understanding concerning environmental requirements such as oxygen availability, water inundation duration and temperature variability, fertilizer requirements for wetland microbes, and characterization of microbes capable of degrading petroleum hydrocarbons.

The overall aim is therefore to advance understanding of the application of constructed treatment wetlands for benzene removal. The objectives are to assess:

- the current literature on benzene removal with constructed wetland systems;
- investigation of the main benzene removal pathways;
- variables and boundary conditions impacting operation and treatment performance (*e.g.*, temperature level and variability, macrophytes, and aggregates);
- the efficiency of different wetland set-ups in removing benzene, COD, BOD, and nutrients;
- the effect of nutrient concentration increases on benzene removal within wetlands; and
- the impact of seasonal change and environmental control on the treatment efficiency of benzene and other water quality variables such as COD, BOD, DO, redox potential, turbidity, and nutrients.

2.6.2   Materials and Methods

2.6.2.1 Experimental System Design and Operation

Two experimental small-scale constructed wetland rigs with six wetlands each are located on The King's Buildings campus in Edinburgh, UK (Table 2.1; Figures 2.27 and 2.28). The systems have been in operation since April 2005. The wetlands were designed to simulate physical, chemical, and microbiological processes occurring in full-scale semi-natural wetlands. One experimental rig was situated outdoors to assess seasonal changes, while the other system was placed indoors to allow better control over the environmental changes. The word 'control' indicates in this case that the data variability of most environmental variables was relatively low due to the relative absence of seasonal impacts.

Since summer 2006, a temperature and humidity control unit was in operation to reduce the variability of temperature even further. The indoor rig was located below three plant grow lights (Sylvania 15000 Hour, 36 W, 1200 mm, T8 Grolux Fluorescent Tube; supplied by Lyco Direct Limited (Bletchely, Milton Keynes, England), product code number: EV1768) simulating day and night conditions.

The wetland rigs were designed to optimize the chemical, physical, and microbiological processes naturally occurring within wetlands. Phenomena studied include biomass manipulations, loading rate variations, and changes to the cycle of filling and emptying of the wetlands.

Round grey polyvinyl chloride drainage pipes (height: 75 cm; diameter: 10 cm), which are resistant to hydrocarbons, were used to construct vertical-flow wetlands. The outlet valves are located at the center of the bottom plate of each wetland and are used for the regulation of flow and sampling. Passive aeration is encouraged with 1.3-cm-diameter ventilation pipes reaching down to 10 cm above the bottom of each wetland. The wetlands are very small in comparison to large-scale systems used in industry, but previous findings based on similar column experiments have been fully accepted by the scientific community (Eke and Scholz 2006; Hiegel 2004; Scholz 2004, 2006a; Omari 2003; Wallace and Knight 2006).

Table 2.1 Packing order of the experimental constructed wetland set-up for inside and outside wetlands. All wetlands were alternatingly inundated and subsequently fully drained two times per week

Height (cm)	Wetland 1	Wetland 2	Wetland 3	Wetland 4	Wetland 5	Wetland 6
61–75 (top)	W+B+F	W+F	W+B+F	W+F	W+B+F	W+F
56–60	5+P+W+B+F	5+P+W+F	5+W+B+F	5+W+F	W+B+F	W+F
51–55	5+P+W+B+F	5+P+W+F	5+W+B+F	5+W+F	W+B+F	W+F
36–50	4+P+W+B+F	4+P+W+F	4+W+B+F	4+W+F	W+B+F	W+F
26–35	3+W+B+F	3+W+F	3+W+B+F	3+W+F	W+B+F	W+F
11–25	2+W+B+F	2+W+F	2+W+B+F	2+W+F	W+B+F	W+F
0–10 (bottom)	1+W+B+F	1+W+F	1+W+B+F	1+W+F	W+B+F	W+F

W: water; B: benzene; F: fertilizer (8 g of N-P-K Miracle-Gro fertilizer were added to all wetlands every 2 weeks until 29 May 2006 when the concentration was increased to 30 g. From 26 June 2006 onwards, the concentration was lowered to 15 g every 2 weeks.); P: *Phragmites australis* (Cav.) Trin. ex Steud. (nine plants of roughly equal biomass and strength per wetland); 1: stones (37.5–75 mm); 2: large gravel (10–20 mm); 3: medium gravel (5–10 mm); 4: small gravel (1.2–5 mm); 5: sand (0.6–1.2 mm).

The wetland systems have different wetland bed volumes depending on the composition of layers of aggregates such as stones, gravel, and sand. Selected wetlands were planted with *Phragmites australis* (Cav.) Trin. ex Steud. (Table 2.1). Nine plants of roughly equal biomass and strength obtained from a local supplier (Alba Trees Public, Lower Winton, Gladsmuir, East Lothian, Scotland) were planted in each 'planted wetland'.

Wetlands 5 and 6 are strictly speaking rather pond systems (extended storage) considering that they do not contain any aggregates. Moreover, wetland 6, con-

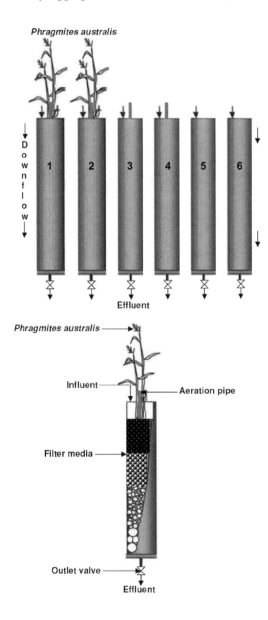

Figure 2.27 Schematic representation showing wetland set-up and internal structure of experimental constructed treatment wetland cell 1

taining only water and fertilizer, can be considered a 'blank.' Wetlands 1, 3, and 5 of both rigs received tap water mixed with a concentration of 1 g/l benzene two times per week. Benzene was used as an example volatile hydrocarbon to assess the removal of low-molecular-weight petroleum compounds. Benzene (BDH analytical reagent, C_6H_6 (99.7%)), supplied by VWR International Ltd (Hunter Boulevard, Lutterworth, UK) was used. Benzene was chosen for various reasons:

- It is a common constituent of liquid fuels.
- It can be used as a surrogate for a mixture of hydrocarbons to allow for easy interpretation of the data and subsequent modeling.
- The traditional treatment technologies used by the oil industry such as hydrocyclones and separators predominantly remove heavy hydrocarbons but not aromatic components in the dissolved water phase.
- The thermodynamic stability of the benzene ring increases its persistence in the environment; therefore, many aromatic compounds are major environmental pollutants.

While hydrocarbons dissolved in water are a suitable source of carbon and energy for microbes, they are an 'incomplete food source' since they do not contain nutrients such as nitrogen, phosphorus, and potassium required for microbial growth. In order to stimulate the growth of microbiological biomass and *P. australis*, 8 g of the well-balanced nitrogen–phosphorus–potassium Miracle-Gro (formerly Osmocote, produced by Scot Europe B. V., The Netherlands) fertilizer were added to all wetlands every 2 weeks until 29 May 2006, when the concentra-

Figure 2.28 Indoor experimental wetland filters treating benzene

tion was increased to 30 g to assess the effect of nutrient concentration increases on benzene removal. From 26 June 2006 onwards, the concentration was lowered to 15 g every 2 weeks, which was seen as a more realistic fertilizer concentration to reduce the fertilizer budget and to avoid too high outflow concentrations, which would have been detrimental to the environment (Table 2.1). The system was designed to operate in batch flow mode to reduce pumping costs.

All wetlands were alternatingly inundated and subsequently fully drained two times per week to encourage rapid air penetration through the soil (Table 2.1) (Cooper *et al.* 1996). During draining, the water flowed rapidly out of the system, and air was subsequently drawn deeply into the lower part of the wetland. This aerated the system and speeded up the rate of hydrocarbon degradation during the 2 or 3 d when the wetlands were not inundated.

Water samples were tested twice per week for COD, pH, DO, turbidity, redox potential, conductivity, ammonia–nitrogen, nitrate–nitrogen, ortho-phosphate–phosphorus, and temperature. The BOD was analyzed in all water samples using a respirometric method with the help of the OxiTOP IS 12-6 system, supplied by the Wissenschaftlich-Technische Werkstatten (WTW), Weilheim, Germany. This equipment uses the principle of piezoresistive measure of pressure differences. Samples were analyzed for benzene removal once per month until January 2007. Samples were tested twice per month afterwards. Benzene content was determined with the Perkin Elmer GC-FID and headspace sampler (models 9700 and HS-101 respectively) instrument.

All water quality variables were determined according to the American Standard Methods for the Examination of Water and Wastewater (1998) unless stated otherwise. Statistical differences were assessed by ANOVA and Tukey's Honestly Significantly Different test ($p < 0.05$), which is based on the "studentized range distribution" (Rice Virtual Lab, Lane D; http://davidmlane.com/hyperstat/B95118.html).

2.6.2.2 Biodegradation and Volatilization Determination

Biodegradation and volatilization were also tested in separate experiments. Two extra wetlands (heights: 24 cm; diameters: 5 cm) were set up under controlled environmental conditions; one wetland comprised aggregates and detritus containing mature microbial biomass (284 g detritus was taken from the upper layer of the contaminated parent wetland 3 located indoors) and the other wetland was left empty. The small wetlands were constructed in the same way as the large wetlands except for the absence of ventilation pipes. The purpose of this auxiliary experiment was to assess the main removal pathways of benzene (combined biodegradation and adsorption versus volatilization) in constructed treatment wetlands. Samples were taken after 1, 2, 3, 6, and 9 d, and benzene content was subsequently determined using headspace and gas chromatography.

2.6.3 Results and Discussion

2.6.3.1 Treatment Performance Comparisons

Considering the benzene inflow concentration of approx. 1.3 g/l, the findings indicate high percentage mean removal efficiencies (85 to 95%) for wetlands containing *P. australis* and aggregates (wetlands 1 and 3 of the indoor and outdoor rig; Table 2.1) and are comparable to data published elsewhere (Myers and Jackson 2001). The findings indicate a high variability of the benzene outflow concentrations. In contrast, wetlands without aggregates and *P. australis* (*i.e.*, wetland 5 of the indoor and outdoor rig; Table 2.1) had slightly lower mean treatment performances (86 to 93%). The removal rate achieved in wetland 5 has been predominantly attributed to the presence of fertilizer (Table 2.1) enhancing the biodegradation rate, because some microbial communities are able to utilize the nitrogen component (*i.e.*, nitrate–nitrogen) of the fertilizer.

The relatively low removal rate for wetland 5 is an indication that aggregates play an indirect role in the treatment of benzene by providing habitat for nitrifying microorganisms (Eke and Scholz 2008). In comparison, the treatment efficiencies for wetlands planted with *P. australis* were similar to the efficiency of the corresponding unplanted wetlands. The treatment performances reduced during the winter of the second year (2006/7) with increasing hydrocarbon accumulation within the corresponding wetlands. This finding is in accordance with data presented previously (Cooper *et al.* 1996; Scholz 2006a).

Concentrations of DO were variable (0 to 12 mg/l). The low DO concentrations (0 to 1 mg/l) in the system during inundation could be due to the lasting effects of rapid aerobic degradation processes observed after drawing air into the system, thus causing a sudden change of the microbiology and subsequent imbalance promoting predominantly aerobic microorganisms in the wetlands. High DO variability was expected considering the relatively high influent concentrations of benzene, low retention time, small size of the experimental wetlands, and sudden environmental changes due to rapid operational changes (see above and Table 2.1).

Control of the variability of environmental variables such as temperature, pressure, and light resulted in improved overall treatment performances of the wetlands. The temperature data (standard deviation: 3.7°C) of the indoor rig were relatively stable, particularly after the temperature was fully controlled (Figure 2.29). Benzene treatment performances were better for the indoor rig versus the outdoor rig. The overall removal efficiencies were lower for the outdoor experimental rig (*e.g.*, benzene, 85 to 86%) in comparison to the experimental rig placed indoors (*e.g.*, benzene, 93 to 95%).

2.6.3.2 Impact of Volatilization

Figure 2.30 shows a comparison of benzene removal for wetlands with and without biomass. The impacts of volatilization, biodegradation, and adsorption on the benzene removal efficiency are often difficult to separate quantitatively from each other. Preliminary findings indicate that biodegradation and adsorption dominate for a retention time of 1 d. However, volatilization becomes the major removal mechanism afterwards. Water and oil are likely to separate if the inflow is not in motion (Eke and Scholz 2008).

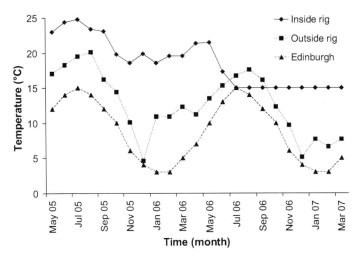

Figure 2.29 Comparison of monthly mean temperatures for inside and outside rigs, and Edinburgh (Met Office (http://www.metoffice.gov.uk/education/secondary/teachers/ukclimate.html#3.2))

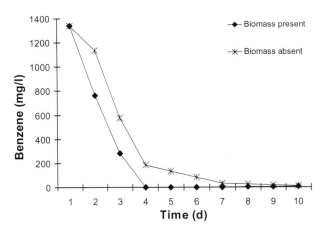

Figure 2.30 Comparison of benzene removal for wetlands with and without biomass

2.6.4   Conclusions

The findings suggest that intermittently flooded vertical-flow constructed wetlands treat benzene effectively in the presence of sufficient oxygen and fertilizer, which provides nitrate as used as an alternative electron acceptor during anaerobic periods of full inundation. Relatively high and stable temperatures (*i.e.*, no seasonal variations) lead to improved hydrocarbon treatment efficiencies.

As benzene and its degradation products started to accumulate in the wetlands, removal efficiencies subsequently diminished. Findings show also that benzene removal was highest in wetlands with aggregates and biomass providing habitat for hydrocarbon-degrading microbes. However, further studies on estimating the microbial biomass are encouraged.

Metabolic processes of microorganisms are likely to play an important role in removing hydrocarbon compounds in both controlled and semi-natural wetlands. The results show also that *P. australis* does not play a significant role (despite providing additional oxygen via its rhizomes) in removing benzene, unless sufficient nutrients (including fertilizer) are available.

Findings indicate also that both biodegradation and volatilization support treatment. Volatilization is the dominant mechanism for benzene removal after 1 d of retention time. However, optimizing environmental conditions such as locating wetlands in areas with relatively high temperatures enhances the biodegradation rate. Further research is required to quantify volatilization, aerobic and anaerobic biodegradation, adsorption, absorption, mineralization, and other removal mechanisms in large-scale constructed treatment wetlands.

References

Abira MA, van Bruggen JJA, Denny P (2005) Potential of a tropical subsurface constructed wetland to remove phenol from pre-treated pulp and paper mill wastewater. Wat Sci Technol 51:173–176

Akratos CS, Tsihrintzis VA (2007) Effect of temperature, HRT, vegetation and porous media on removal efficiency of pilot-scale horizontal subsurface flow constructed wetlands. Ecol Eng 29:173–191

APHA (1998) Standard methods for the examination of water and wastewater, 20th edn. American Public Health Association (APHA), American Water Works Association and Water and Environmental Federation, Washington, DC

Arimoro FO, Osakwe EI (2006) The influence of sawmill wood wastes on the distribution and population of macroinvertebrates at Benin River, Niger Delta area, Nigeria. Chem Biodiv 3:578–592

Arvidsson S (2006) En ekotoxikologisk studie av lakvatten från timmerbevattning (An ecotoxicological study of runoff from sprinkling of roundwood). Master's thesis, LITH-IFM-EX-05/1508-SE, Linköping University, Linköping, Sweden

Atlas RM, Cerniglia CE (1995) Bioremediation of petroleum pollutants. BioSci 45:332–338

Babatunde AO, Zhao YQ, O'Neill M, O'Sullivan B (2008) Constructed wetlands for environmental pollution control: a review of developments, research and practice in Ireland. Environ Int 34:116–126

Bailey HC, Elphick JR, Potter A, Chao E, Konasewich D, Zak JB (1999) Causes of toxicity in storm water runoff from sawmills. Environ Toxicol Chem 18:1485–1491

Baris AJ, Eifert WH, Klotzer K, McGuckin CJ (2001) Use of a sub-surface flow constructed wetland for collection and treatment of water containing BTEX. Roux Associates, Inslandia, New York

Bastviken SK, Eriksson PG, Premrov A, Tonderski KS (2005) Potential denitrification in wetland sediments with different plant species detritus. Ecol Eng 25:183–190

Bonomo L, Pastorelli G, Zambon N (1997) Advantages and limitations of duckweed-based wastewater treatment systems. Wat Sci Technol 35:239–246

Borgå P, Elowson T, Liukko K (1996a) Environmental loads from water-sprinkled softwood timber: 1. Characteristics of an open and a recycling watering system. Environ Toxicol Chem 15:856–867

Borgå P, Elowson T, Liukko K (1996b) Environmental loads from water-sprinkled softwood timber: 2. Influence of tree species and water characteristics on wastewater discharges. Environ Toxicol Chem 15:1445–1454

Bowmer KH, Laut P (1992) Waste-water management and resource recovery in intensive rural industries in Australia. Wat Res 26:201–208

Braskerud BC (2002) Factors affecting phosphorus retention in small constructed wetlands treating agricultural non-point source pollution. Ecol Eng 1:41–61

Brix H (1994) Functions of macrophytes in constructed wetlands. Wat Sci Technol 29:71–78

Brix H, Schierup H-H, Arias CA (2007) Twenty years experience with constructed wetland systems in Denmark – what did we learn? Wat Sci Technol 56:63–68

Burland SB, Edwards EA (1999) Anaerobic benzene biodegradation linked to nitrate reduction. Appl Environ Microbiol 65:529–533

Buss SR, Herbert AW, Morgan P, Thorton SF, Smith JWN (2004) A review of ammonium attenuation in soil and groundwater. Q J Eng Geo Hydrogeol 37:347–359

Cameron K, Madramootoo C, Crolla A, Kinsley C (2003) Pollutant removal from municipal sewage lagoon effluents with a free-surface wetland. Wat Res 37:2803–2812

Carmichael LM, Pfaender FK (1997) The effect of inorganic and organic supplements on the microbial degradation of phenanthrene and pyrene in soils. Biodegradation 8:1–13

Carty A, Scholz M, Heal K, Gouriveau F, Mustafa A (2008) The universal design, operation and maintenance guidelines for farm constructed wetlands (FCW) in temperate climates. Biores Technol 99:6780–6792

Carroll P, Harrington R, Keohane J, Ryder C (2005) Water treatment performance and environmental impact of integrated constructed wetlands in the Anne valley watershed, Ireland. In: Dunne EJ, Reddy KR, Carton OT (eds) Nutrient management in agricultural watersheds: a wetlands solution. Wageningen Academic Publishers, Wageningen, The Netherlands, pp. 207–217

Casson JP, Bennett DR, Nolan SC, Olson BM, Ontkean GR (2006) Degree of phosphorus saturation thresholds in manure-amended soils of Alberta. J Environ Qual 35:2212–2221

Caswell PC, Gelb D, Marinello SA, Emerick JC, Cohen RR (1992) Evaluation of constructed surface-flow wetlands systems for the treatment of discharged waters from oil and gas operations in Wyoming. In: Proceedings of the SPE Rocky Mountain regional conference, Paper SPE 24331, Casper, WY

Chaillan F, Chaineau CH, Point V, Saliot A, Oudot J (2006) Factors inhibiting bioremediation of soil contaminated with weathered oils and drill cuttings. Environ Pollut 144: 255–265

Chaineau CH, Rougeux G, Yepremian C, Oudot J (2005) Effects of nutrient concentration on the biodegradation of crude oil and associated microbial populations in the soil. Soil Biol Biochem 37:1490–1497

Chalkias C, Petrakis M, Psiloglou B, Lianou M (2006) Modelling of light pollution in suburban areas using remotely sensed imagery and GIS. J Environ Manag 79:57–63

Chen SW, Kao CM, Jou CR, Fu YT, Chang YI (2008) Use of a constructed wetland for post-treatment of swine wastewater. Environ Eng Sci 25:407–417

Choi S-C, Kwon KK, Sohn JH, Kim S-J (2000) Evaluation of fertilizer additions to stimulate oil biodegradation in sand seashore mescocosms. J Microbiol Biotechnol 12:431–436

Ciria MP, Solano ML, Soriano P (2005) Role of macrophyte *Typha latifolia* in a constructed wetland for wastewater treatment and assessment of its potential as biomass fuel. Biosyst Eng 92:535–544

Clarke E, Baldwin AH (2002) Responses of wetland plants to ammonia and water level. Ecol Eng 18:257–264

Cleneghan C (2003) Phosphorus regulations; national implementation report. Environmental Protection Agency, Wexford, Ireland

Coates JD, Chakraborty R, McInerney MJ (2002) Anaerobic benzene biodegradation – a new era. Res Microbiol 153:621–628

Conte G, Martinuzzi N, Giovannelli L, Pucci B, Masi F (2001) Constructed wetlands for wastewater treatment in central Italy. Wat Sci Technol 44:339–343

Cook MG, Hunt PG, Stone KC, Canterberry JH (1996) Reducing diffuse pollution through implementation of agricultural best management practices: a case study. Wat Sci Technol 33:191–196

Cooper PF, Job GD, Green MB, Shutes RBE (1996) Reed beds and constructed wetlands for wastewater treatment. Water Research Centre, Swindon, UK

Cronk JK (1996) Constructed wetlands to treat wastewater from dairy and swine operations: a review. Agric Ecosyst Environ 58:97–114

Dagley S (1986) Biochemistry of aromatic hydrocarbon degradation in *Pseudomonas*. In: Sokatch J, Ornston LN (eds) The bacteria. Academic, Orlando, FL, USA, vol 10, pp. 527–555

Daly K (2000) Phosphorus desorbtion from Irish soils. In: Tunney H (ed) Quantification of phosphorus loss from soil to water. Environmental Protection Agency, Wexford, UK, pp. 59–101

DeHoop CF, Einsel DA, Ro KS, Chen S, Gibson MD, Grozdits GA (1998) Storm water runoff quality of a Louisiana log storage and handling facility. J Environ Sci Health A Toxic Hazard Subst Environ Eng 33:165–177

Di Lecce C, Accarino M, Bolognese F, Galli E, Barbieri P (1997) Isolation and metabolic characterization of a *Pseudomonas stutzeri* mutant able to grow on the three isomers of xylene. Appl Environ Microbiol 63:3279–3281

Díaz E (2004) Bacterial degradation of aromatic pollutants: a paradigm of metabolic versatility. Int Microbiol 7:173–180

Doig P, van Poppelen P, Baldwin SA (2006) Characterization of particles in fresh and primary-treated log sort yard runoff. Wat Qual Res J Can 41:37–46

Doradoa J, Claassena FW, Lenonc G, van Beek TA, Wijnberg JBPA, Sierra-Alvareza R (2000) Degradation and detoxification of softwood extractives by sapstain fungi. Biores Technol 71:13–20

Drizo A, Comeau Y, Forget C, Chapuis RP (2002) Phosphorus saturation potential: a parameter for estimating the longevity of constructed wetland systems. Environ Sci Technol 36:4642–4648

Dunne EJ, Culleton N, O'Donovan G, Harrington R, Olsen AE (2005a) An integrated constructed wetland to treat contaminants and nutrients from dairy farmyard dirty water. Ecol Eng 24:221–234

Dunne EJ, Culleton N, O'Donovan G, Harrington R, Daly K (2005b) Phosphorus retention and sorption by constructed wetland soils in Southeast Ireland. Wat Res 39:4355–4362

EEC (1991) European Economic Community (EEC) Council Directive 91/676/EEC of 12 December 1991. Concerning the protection of waters against pollution caused by nitrates from agricultural sources. Official J L 375, 1–8

Eke PE, Scholz M (2006) Hydrocarbon removal with constructed treatment wetlands for the benefit of the petroleum industry. In: Dias V, Vymazal J (eds) 10th international conference on wetland systems for water pollution control (23-29/09/2006). International Water Association, Lisbon, Portugal, 3:1707–1714

Eke PE, Scholz M (2008) Benzene removal with vertical-flow constructed treatment wetlands. J Chem Technol Biotechnol 83:55–63

El-Din MG, Smith DW, Al Momani F, Ikehata K (2006a) Ozone treatment for the degradation of resin and unsaturated fatty acids at low temperatures. J Environ Eng Sci 5:95–102

El-Din MG, Smith DW, Al Momani F, Wang W (2006b) Oxidation of resin and fatty acids by ozone: kinetics and toxicity study. Wat Res 40:392–400

Environmental Protection Agency (1998) Irish Phosphorus Regulations 1998. Water quality standards for phosphorus. Local Government (Water Pollution) Act 1977. Irish Government, Dublin, Ireland. S.I. No. 258 1998: 1998;
http://www.epa.ie/whatwedo/enforce/pa/phosphorus/. Accessed 15 Jun 2009

EPA (2002) Interim report: the biological survey of river quality 2001. Environmental Protection Agency, Johnston Castle, Wexford, UK

Eriksson PG, Weisner SEB (1996) Functional differences in epiphytic microbial communities in nutrient-rich freshwater ecosystems: an assay of denitrifying capacity. Freshwat Biol 36: 552–562

Erskine AD (2000) Transport of ammonium in aquifers: retardation and degradation. Q J Eng Geo Hydrogeol 33:161–170

Field JA, Leyendeckers, MJH, Alvarez RS, Lettinga G, Habets LHA (1988) The methanogenic toxicity of bark tannins and the anaerobic biodegradability of water-soluble bark matter. Wat Sci Technol 20:219–240

Feist WC, Springer EL, Hajny GJ (1971) Viability of parenchyma cells in stored green wood. Tech Assoc Pulp Paper Ind (TAPPI, now TAPPI Journal) 54:1295–1297

Froneman A, Mangnall MJ, Little RM, Crowe TM (2001) Water bird assemblages and associated habitat characteristics of farm ponds in the Western Cape, South Africa. Biodiv Conserv 10:251–270

Gottschall N, Boutin C, Crolla A, Kinsley C, Champagne P (2007) The role of plants in the removal of nutrients at a constructed wetland treating agricultural (dairy) wastewater, Ontario, Canada. Ecol Eng 29:154–163

Guirguis M (2004) Treatment of waste water: a reed bed environmental case history. In: SPE International conference on health, safety, and environment in oil and gas exploration and production, Paper SPE86673, Calgary, Alberta, Canada, 29–31 March 2004

Haberl H, Erb KH, Krausmann F, Adensam H, Schulz N (2003) Land-use change and socio-economic metabolism in Austria: II. Land-use scenarios for 2020. Land Use Policy 20:21–39

Hamouri B, El Nazih J, Lahjouj J (2007) Subsurface-horizontal flow constructed wetland for sewage treatment under Moroccan climate conditions. Desalination 215:153–58

Harayama S, Kasai Y, Hara AA (2004) Microbial communities in oil contaminated seawater. Curr Opin Biotechnol 15:205–214

Harrington A (2005) The relationship between plant vigour and ammonium concentrations in surface waters of constructed wetlands used to treat meat industry wastewaters in Ireland. In: Dunne EL, Reddy KR, Carton OT (eds) Nutrient management in agricultural watersheds: a wetlands solution. Wageningen Academic Publishers, Wageningen, The Netherlands, pp. 219–223

Harrington R, Ryder C (2002) The use of integrated constructed wetlands in the management of farmyard runoff and waste water. In: Proceedings of the national hydrology seminar on water resource management: sustainable supply and demand, Irish National Committees of the International Hydrological Programme and International Commission on Irrigation and Drainage. Tullamore, Offaly, UK, pp. 55–63

Harrington R, Dunne EJ Carroll P, Keohane J, Ryder C (2005) The concept, design and performance of integrated constructed wetlands for the treatment of farmyard dirty water. In: Dunne EL, Reddy KR, Carton OT (eds) Nutrient management in agricultural watersheds: a wetlands solution. Wageningen Academic Publishers, Wageningen, The Netherlands, pp. 179–188

Harrington R, Carroll P, Carty A, Keohane J, Ryder C (2007) Integrated constructed wetlands: concept, design, site evaluation and performance. Int J Wat 3:243–256

Harrison JJ, Turner RJ, Marques LLR, Ceri H (2005) Biofilms: a new understanding. Am Sci 93(6) 508–515

He LS, Liu HL, Xi BD, Zhu YB (2006) Enhancing treatment efficiency of swine wastewater by effluent recirculation in vertical-flow constructed wetland. J Environ Sci China 18:221–226

Head IM, Saunders JR, Pickup RW (1998) Microbial evolution, diversity, and ecology; a decade of ribosomal RNA analysis of uncultured microorganisms. Microbiol Ecol 35:1–21

Healy MG, Rodgers M, Mulqueen J (2007) Treatment of dairy wastewater using constructed wetlands and intermittent sand filters. Biores Technol 98:2268–2281

Hedmark Å, Jonsson M (2008) Treatment of log yard runoff in a couch grass infiltration wetland in Sweden. Int J Environ Stud 65:273–278

Hedmark Å, Scholz M (2008) Review of environmental effects and treatment of runoff from storage and handling of wood. Biores Technol 99:5997–6009

Hefting MM, Bobbink R, de Caluwe H (2003) Nitrous oxide emission and denitrification in chronically nitrate-loaded riparian buffer zones. J Environ Qual 32:1194–1203

Heider J, Spormann AM, Beller HR, Widdel F (1998) Anaerobic bacterial metabolism of hydrocarbons. FEMS Microbiol Rev 22:459–473

Hiegel T (2004) Analysis of pilot scale constructed wetland treatment of petroleum contaminated groundwater. MSc thesis, Department of Civil Engineering, University of Wyoming, Laramie, WY, USA

Hill VR, Sobsey MD (2001) Removal of Salmonella and microbial indicators in constructed wetlands treating swine wastewater. Wat Sci Technol 44:215–222

Hill DT, Payne VWE, Rogers JW, Kown SR (1997) Ammonia effects on the biomass production of five constructed wetland plant species. Biores Technol 62:109–113

Hill CM, Duxbury J, Geohring L, Peck T (2000) Designing constructed wetlands to remove phosphorus from barnyard runoff: a comparison of four alternative substrates. J Environ Sci Health A Toxic Hazard Subst Environ Eng 35:1357–1375

Hilton J (2003) Reducing diffuse pollution to rivers: a dictionary of Best Management Practices. Conserving Natura 2000 Rivers Conservation Techniques Series No. 10, English Nature, Peterborough, UK

Hubbard RK, Ruter JM, Newton GL, David JG (1999) Nutrient uptake and growth response of six wetland/riparian plant species receiving swine lagoon effluent. Trans Am Soc Agric Eng 42:1331–1341

Hubbard RK, Gascho GJ, Newton GL (2004) Use of floating vegetation to remove nutrients from swine lagoon wastewater. Trans Am Soc Agric Eng 47:1963–1972

Hunt PG, Szogi AA, Humenik FJ, Rice JM, Matheny TA, Stone KC (2002) Constructed wetlands for treatment of swine wastewater from an anaerobic lagoon. Trans Am Soc Agric Eng 45:639–647

Hunt PG, Matheny TA, Stone KC (2004) Denitrification in a coastal plain riparian zone contiguous to a heavily loaded swine wastewater spray field. J Environ Qual 33:2367–2374

Humenik FJ, Szogi AA, Hunt PG, Broome S, Rice M (1999) Wastewater utilization: a place for managed wetlands – review. Asian-Australas J Animal Sci 12:629–632

Ibekwe AM, Grieve CM, Lyon SR (2003) Characterization of microbial communities and composition in constructed dairy wetland wastewater effluent. Appl Environ Microbiol 69: 5060–5069

Irish Phosphorous Regulations (1998) Water quality standards for phosphorous. Local Government (Water Pollution) Act 1977. S.I. No. 258 1998. Irish Government, Dublin, Ireland

Janks JS, Cadena F (1991) Identification and properties of modified zeolites for the removal of benzene, toluene and xylene from aqueous solutions, paper SPE 22833. In: 1991 Society of Petroleum Engineers (SPE) annual technical conference and exhibition, Dallas, TX, 6–9 October 1991

Jiang CL, Fan XQ, Cui GB, Zhang YB (2007) Removal of agricultural non-point source pollutants by ditch wetlands: implications for lake eutrophication control. Hydrobiology 581: 319–327

Jonsson M (2004) Wet storage of round wood. PhD thesis. Acta Universitatis Agriculturae Sueciae – Silvestria, vol 319. Swedish University of Agricultural Sciences, Uppsala, Sweden

Jonsson M, Dimitriou I, Aronsson P, Elowson T (2004) Effects of soil type, irrigation volume and plant species on treatment of log yard run-off in lysimeters. Wat Res 38:3634–3642

Jonsson M, Dimitriou I, Aronsson P, Elowson T (2006) Treatment of log yard runoff by irrigation of grass and willows. Environ Pollut 139:157–166

Kadlec RH (1999) The limits of phosphorus removal in wetlands. Wetland Ecol Manag 7: 165–175

Kadlec RH (2001) Thermal environments of subsurface treatment wetlands. Wat Sci Tech 44:251–258

Kadlec RH, Knight RL (1996) Treatment wetlands. CRC, Boca Raton, FL

Kadlec RH, Knight RL, Vymazal J, Brix H, Cooper P, Haberl R (2000) Constructed wetlands for pollution control. Scientific and technical report No. 8. International Water Association Publishing, London, UK

Kantawanichkul S, Somprasert S (2005) Using a compact combined constructed wetland system to treat agricultural wastewater with high nitrogen. Wat Sci Technol 51:47–53

Kantawanichkul S, Neamkam P, Shutes RBE (2001) Nitrogen removal in a combined system: vertical vegetated bed over horizontal sand bed. Wat Sci Technol 44:137–142

Kantawanichkul S, Somprasert S, Aekasin U, Shutes RBE (2003) Treatment of agricultural wastewater in two experimental combined constructed wetland systems in a tropical climate. Wat Sci Technol 48:199–205

Karim MR, Manshadi FD, Karpiscak MM, Gerba CP (2004) The persistence and removal of enteric pathogens in constructed wetlands. Wat Res 38:1831–1837

Karrasch B, Parra O, Cid H, Mehrens M, Pacheco P, Urrutia R, Valdovinos C, Zaror C (2006) Effects of pulp and paper mill effluents on the microplankton and microbial self-purification capabilities of the Biobio River, Chile. Sci Total Environ 359:194–208

Kauppinen T, Vincent R, Liukkonen T, Grzebyk M, Kauppinen A, Welling I, Arezes P, Blacks N, Bochmann F, Campelo F, Costa M, Elsigan G, Goerens R, Kikemenis A, Kromhout H, Miguel S, Mirabelli D, McEneany R, Pesch B, Plato N, Schlunssen V, Schulze J, Sonntag R, Verougstraete V, De Vincente MA, Wolf J, Zimmermann M, Husgafvel-Pursiainen K, Savolainen K (2006) Occupational exposure to inhalable wood dust in the member states of the European Union. Ann Occupat Hyg 50:549–561

Kayranli B, Scholz M, Mustafa A, Hofmann O, Harrington R (2010) Performance evaluation of integrated constructed wetlands treating domestic wastewater. Wat Air Soil Pollut 210:435–451

Kellner E, Price S, Waddington JM (2004) Pressure variations in peat as a result of gas bubble dynamics. Hydrol Proc 18:2599–2605

Kent DM (2000) Applied wetlands science and technology, 2nd edn. CRC, Boca Raton, FL

Keohane J, Carroll P, Harrington R, Ryder C (2005) Integrated constructed Wetlands for farmyard dirty water treatment: a site suitability assessment. In: Dunne EJ, Reddy KR, Carton OT (eds) Nutrient management in agricultural watersheds: a wetlands solution. Wageningen Academic Publishers, Wageningen, The Netherlands, p 196–206

Kim S, Choi DH, Sim DS, Oh Y (2005) Evaluation of bioremediation effectiveness on crude oil-contaminated sand. Chemosphere 59:845–852

Kivaisi AK (2001) The potential for constructed wetlands for wastewater treatment and reuse in developing countries: a review. Ecol Eng 16:545–560

Knight RL (1999) The use of treatment wetlands for petroleum industry effluents. Environ Sci Technol 33:973–980

Knight RL, Payne Jr WE, Borer RE, Clarke Jr RA, Pries JH (2000) Constructed wetlands for livestock wastewater management. Ecol Eng 15:41–55

Kowalchuk GA, Stephen JR, De Boer W, Prosser JI, Embley TM, Woldendorp JW (1997) Analysis of ammonia-oxidizing bacteria of the β subdivision of the class Proteobacteria in coastal sand dunes by denaturing gradient gel electrophoresis and sequencing PCR-amplified 16S ribosomal DNA fragments. Appl Environ Microbiol 63:1489–1497

Kraus H (1985) Noise abatement in sawmills. Holz-Zentralblatt 111:1833–1834

Lahvis MA, Baehr AL, Baker RJ (1999) Quantification of aerobic biodegradation and volatilization rates of gasoline hydrocarbons near the water table under natural attenuation conditions. Wat Resour Res 35:753–765

Lansing SL, Martin JF (2006) Use of an ecological treatment system (ETS) for removal of nutrients and solids from dairy wastewater. Ecol Eng 28:235–245

Leahy JG, Olsen RH (1997) Kinetics of toluene degradation by toluene-oxidizing bacteria as a function of oxygen concentration, and the effect of nitrate. FEMS Microbiol Ecol 23:23–30

Lee CY, Lee CC, Lee FY, Tseng SK, Laio CJ (2004) Performance of subsurface flow constructed wetland taking pre-treated swine effluent under heavy loads. Biores Technol 92:173–179

Lee B-H, Scholz M, Horn A (2006) Constructed wetlands for the treatment of concentrated stormwater runoff (Part A). Environ Eng Sci 23:191–202

Leinweber P, Meissner R, Eckhardt K-U, Seeger J (1999) Management effects on forms of phosphorus in soil and leaching losses. Eur J Soil Sci 50:413–424

Li L, Li Y, Kumar Biswas D, Nian Y, Jiang G (2008) Potential of constructed wetlands in treating the eutrophic water: evidence from Taihu Lake of China. Biores Technol 99:1656–1663

Lindholm EL, Berg S (2005) Energy requirement and environmental impact in timber transport. Scand J For Res 20:184–191

Liukko K. (1997) Climate-adapted wet storage of saw timber and pulpwood: An alternative method of sprinkling and its effect on freshness of roundwood and environment. PhD thesis. Acta Universitatis Agriculturae Sueciae – Silvestria, No. 51. Swedish University of Agricultural Sciences, Uppsala, Sweden

Liukko K, Elowson T (1995) Climate controlled sprinkling of saw timber. In: Web Proceedings of the International Union of Forestry Research Organizations (IUFRO) XX World Congress, P3.07 Meeting, Tampere, Finland, 6–12 August, pp. 44–53. Oregon State University, Corvallis, OR, USA (http://www.metla.fi/iufro/iufro95/index.htm)

Liukko K, Elowsson T (1999) The effect of bark condition, delivery time and climate-adapted wet storage on the moisture content of *Picea abies* (L.) Karst. pulpwood. Scand J For Res 14:156–163

Lovley DR, Lloyd JR (2000) Microbes with a mettle for bioremediation. Nat Biotechnol 18:600–601

Machate T, Noll BHH, Kettrup A (1997) Degradation of Phenanthrene and hydraulic characteristics in a constructed wetland. Wat Res 31:554–560

Maguire RO, Edwards AC, Wilson MJ (1997) Comparison of chemical forms and distribution of phosphorus within cultivated and uncultivated soils; some implications for losses. In: Tunney H, Carton OT, Brooks PC, Johnson AE (eds) Phosphorus losses from soil to water. CAB Wallingford, UK, pp. 427–430

Mantovi P, Marmiroli M, Maestri E, Tagliavini S, Piccinni S, Marmiroli N (2003) Application of a horizontal subsurface flow constructed wetland on treatment of dairy parlor wastewater. Biores Technol 88:85–94

Masbough A, Frankowski K, Hall KJ, Duff SJB (2005) The effectiveness of constructed wetland for treatment of woodwaste leachate. Ecol Eng 25:552–566

McCuskey SA, Conger AW, Hillestad HO (1994) Design and implementation of functional wetland mitigation – case studies in Ohio and South Carolina. Wat Air Soil Pollut 77:513–532

McMaster ME, Hewitt LM, Parrott JL (2006) A decade of research on the environmental impacts of pulp and paper mill effluents in Canada: field studies and mechanistic research. J Toxicol Environ Health B Crit Rev 9:319–339

Met Éireann (2007) http://www.met.ie/climate/climate-data-information.asp. Accessed 15 Jan 2010

Mitsch WJ, Gosselink JG (1993) Wetlands. 2nd edn. Wiley, New York

Mitsch WJ, Gosselink JG (2007) Wetlands. 4th edn. Wiley, New York

Molle P, Prost-Boucle S, Lienard A (2008) Potential for total nitrogen removal by combining vertical flow and horizontal flow constructed wetlands: a full-scale experiment study. Ecol Eng 34:23–29

Moshiri GA (2000) Constructed wetland for water quality improvement. CRC, Boca Raton, FL, USA

Munoz I, Rieradevall J, Torrades F, Peral J, Domenech X (2006) Environmental assessment of different advanced oxidation processes applied to a bleaching Kraft mill effluent. Chemosphere 62:9–16

Mustafa A, Scholz M, Harrington R, Carrol P (2009) Long-term performance of a representative integrated constructed wetland treating farmyard runoff. Ecol Eng 35:779–790

Myers JE, Jackson LM (2001) An evaluation of the Department of Energy Naval Petroleum Reserve No. 3 – Produced water bio-treatment facility. In: Proceedings of SPE/EPA/DOE exploration and production environmental conference, SPE Paper 66522, San Antonio, TX

Nair I, Jayachandran K, Shankar S (2007) Treatment of paper factory effluent using a phenol degrading *Alcaligenes sp.* under free and immobilized conditions. Biores Technol 98: 714–716

Newman JM, Clausen JC, Neafsey JA (1999) Seasonal performance of a wetland constructed to process dairy milkhouse wastewater in Connecticut. Ecol Eng 14:181–198

Nicholson BC (1987) Changes in volatiles composition of Pinus radiata on wet storage. Holzforschung 41:209–213

Nielsen LP, Christensen PB, Revsbech NP, Sørensen J (1990) Denitrification and oxygen respiration in biofilms studied with a microsensor for nitrous oxide and oxygen. Microb Ecol 19:63–72

NIWA (1997) Guidelines for constructed wetland treatment of farm dairy wastewaters in New Zealand. Science and Technology Series No. 48, National Institute of Water and Atmospheric Research Limited (NIWA), http://www.niwascience.co.nz/pubs/st/st48.pdf. Accessed 25 Sep 2007

Omari K (2003) Hydrocarbon removal in an experimental gravel bed constructed wetland. Wat Sci Technol 48:275–281

Orban JL, Kozak RA, Sidle RC, Duff SJB (2002) Assessment of relative environmental risk from log yard runoff in British Columbia. For Chr 78:146–151

Oudot J, Merlin FX, Pinvidic P (1998) Weathering rates of oil components in a bioremediation experiment in estuarine sediments. Mar Environ Res 45:113–125

Paje MLF, Neilan BA, Couperwhite IA (1997) *Rhodococcus* species that thrives on medium saturated with liquid benzene. Microbiology 143:2975–2981

Pant HK, Reddy KR, Lemon E (2001) Phosphorus retention capacity of root bed media of subsurface flow constructed wetlands. Ecol Eng 17:345–355

Pease BC (1974) Effects of log dumping and rafting on the marine environment of southeast Alaska. USDA Forest Service General Technical Report PNW-22, Pacific Northwest Forest and Range Experiment Station, US Department of Agriculture, Forest Service, Portland, OR

Peek RD, Liese W (1974) The effect of wet storage of windthrown timber on water quality. Forstwissenschaftliches Zentralblatt 96:348–357

Picard C, Fraser HL, Steer D (2005) The interacting effects of temperature and plant community type on nutrient removal in wetland microcosms. Biores Technol 96:1039–1047

Picek T, Cızkova H, Dusek J (2007) Greenhouse gas emissions from a constructed wetland – plants as important sources of carbon. Ecol Eng 31:98–106

Poach ME, Hunt PG, Sadler EJ, Matheny TA, Johnson MH, Stone KC, Humenik FJ, Rice JM (2002) Ammonia volatilisation from constructed wetlands that treat swine wastewater. Trans Am Soc Agric Eng 45:619–627

Poach ME, Hunt PG, Vanotti MB, Stone KC, Matheny TA, Johnson MH, Sadler EJ (2003) Improved nitrogen treatment by constructed wetlands receiving partially nitrified liquid swine manure. Ecol Eng 20:183–197

Poach ME, Hunt PG, Reddy GB, Stone KC, Matheny TA, Johnson MH, Sadler EJ (2004) Ammonia volatilization from marsh-pond-marsh constructed wetland treating swine wastewater. J Environ Qual 33:844–851

Poe AC, Pichler MF, Thompson SP, Paerl HW (2003) Denitrification in a constructed wetland receiving agricultural runoff. Wetlands 23:817–826

Prantner SR, Kanwar RS, Lorimor JC, Pederson CH (2001) Soil infiltration and wetland microcosm treatment of liquid swine manure. Appl Eng Agric 17:483–488

Puustinen M, Jormola J (2005) Constructed wetlands for nutrient retention and landscape diversity. International Commission on Irrigation and Drainage 21st European Regional Conference 2005. Frankfurt an der Oder, Germany

Ragusa SR, McNevin D, Qasem S, Mitchell C (2004) Indicators of biofilm development and activity in constructed wetlands microcosms. Wat Res 38:2865–2873

Ran N, Agami M, Oron G (2004) A pilot study of constructed wetlands using duckweed (*Lemna gibba* L.) for treatment domestic primary effluent in Israel. Wat Res 38:2241–2248

Reddy KR, D'Angelo EM (1997) Biogeochemical indicators to evaluate pollutant removal efficiency in constructed wetlands. Wat Sci Technol 35:1–10

Reddy KR, Kadlec RH, Flaig E, Gale PM (1999) Phosphorus retention in streams and wetlands. Crit Rev Environ Sci Technol 29:83–146

Reddy GB, Hunt PG, Phillips R, Stone K, Grubbs A (2001) Treatment of swine wastewater in marsh-pond-marsh constructed wetlands. Wat Sci Technol 44:545–550

Rew S, Mulamoottil G (1999) A cost comparison of leachate treatment alternatives. In: Mulamoottil G, McBean EA, Rovers F (eds) Constructed wetlands for the treatment of landfill leachates. Lewis, Boca Raton, FL, USA

Richardson CJ, Marshall PE (1986) Processes controlling movement, storage, and export of phosphorus in a fen peatland. Ecol Monogr 56:279–302

Rice RW, Erich MS (2006) Estimated VOC losses during the drying of six eastern hardwood species. For Prod J 56:48–51

Rice RW, Izuno FT, Garcia RM (2002) Phosphorus load reductions under best management practices for sugarcane cropping systems in the Everglades Agricultural Area. Agric Wat Manag 56:17–39

Ridgeway HF, Safarik J, Phipps D, Carl P, Clark D (1990) Identification and catabolic activity of well-derived gasoline-degrading bacteria and a contaminated aquifer. Appl Environ Microbiol 56:3565–3575

Rintala JA, Puhakka JA (1994) Anaerobic treatment in pulp- and paper-mill waste management: a review. Biores Technol 47:1–18

Sakadevan K, Bavor H (1998) Phosphate adsorption characteristics of soils, Slags and zeolite to be used as substrates in constructed wetland systems. Wat Res 32:393–399

Samis SC, Liu SD, Wernick BG, Nassichuk MD (1999) Mitigation of fisheries impacts from the use and disposal of wood residue in British Columbia and the Yukon. Canadian Technical Report of Fisheries and Aquatic Sciences 2296. Fisheries and Oceans Canada, and Environment Canada. Ottawa, Ontario, Canada

Schaafsma JA, Baldwin AH, Streb CA (2000) An evaluation of a constructed wetland to treat wastewater from a dairy farm in Maryland, USA. Ecol Eng 14:199–206

Scholz M (2004) Treatment of gully pot effluent containing nickel and copper with constructed wetlands in a cold climate. J Chem Technol Biotechnol 79:153–162

Scholz M (2006a) Wetland systems to control urban runoff. Elsevier, Amsterdam, The Netherlands

Scholz M (2006b) Practical sustainable urban drainage system decision support tools. Inst Civ Eng Eng Sustain 159:117–125

Scholz M (2007) Classification methodology for sustainable flood retention basins. Landsc Urban Plann 81:246–256

Scholz M, Xu J (2001) Comparison of vertical-flow constructed wetlands for treatment of wastewater containing lead and copper. J Chart Inst Wat Environ Manag 15:287–293

Scholz M, Xu J (2002) Performance comparison of experimental constructed wetlands with different filter media and macrophytes treating industrial wastewater contaminated with lead and copper. Biores Technol 83:71–79

Scholz M, Lee B-H (2005) Constructed wetlands: a review. Int J Env Stud 62:421–447

Scholz M, Höhn P, Minall R (2002) Mature experimental constructed wetlands treating urban water receiving high metal loads. Biotechnol Progr 18:1257–1264

Scholz M, Sadowski AJ, Harrington R, Carroll P (2007a) Integrated constructed wetlands assessment and design for phosphate removal. Biosyst Eng 97:415–423

Scholz M, Harrington R, Carroll P, Mustafa A (2007b) The integrated constructed wetlands (ICW) concept. Wetlands 27:337–354

Seo DC, Cho JS, Lee HJ, Heo JS (2005) Phosphorus retention capacity of filter media for estimating the longevity of constructed wetland. Wat Res 39:2445–2457

Sezerino PH, Reginatto V, Santos MA, Kayser K, Kunst S, Philippi LS, Soares HM (2003) Nutrient removal from piggery effluent using vertical flow constructed wetlands in southern Brazil. Wat Sci Technol 48:129–135

Shappell NW, Billey LO, Forbes D, Matheny TA, Poach ME, Reddy GB, Hunt PG (2007) Estrogenic activity and steroid hormones in swine wastewater through a lagoon constructed wetland system. Environ Sci Technol 41:444–450

Sharpley AN (1995) Identifying sites vulnerable to phosphorus losses in agricultural runoff. J Environ Qual 24:947–951

Shipin O, Koottatep T, Khanh NTT, Polprasert C (2005) Integrated natural treatment systems for developing communities: low-tech N-removal through the fluctuating microbial pathways. Wat Sci Technol 51:299–306

Sievers DM (1997) Performance of four constructed wetlands treating anaerobic swine lagoon effluents. Trans Am Soc Agric Eng 40:769–775

Skarda SM, Moore JA, Niswander SF, Gamroth MJ (1994) Preliminary results of a wetland for treatment of dairy farm wastewater. In: DuBowy PJ, Reaves RP (eds) Constructed wetlands for animal wastewater management. Proceedings of workshop, 4–6 Apr 1994, Purdue University, West Lafayette, IN, pp. 34–42

Solano-Serena F, Marchal R, Ropars M, Lebeault J-M, Vandecasteele J-P (1999) Biodegradation of gasoline: kinetics, mass balance and fate of individual hydrocarbons. J Appl Microbiol 86:1008–1016

Solano ML, Soriano P, Ciria MP (2003) Constructed wetlands as a sustainable solution for wastewater treatment in small villages. Biosyst Eng 87:109–118

Sommer SG, Christensen BT, Nielsen NE, Schjorring JK (1993) Ammonia volatilization during storage of cattle and pig slurry – effect of surface cover. J Agric Sci 121:63–71

Soukup A, Williams RJ, Cattell FCR, Krough MH (1994) The function of a coastal wetland as an efficient remover of nutrients from sewage effluent: a case study. Wat Sci Technol 29:295–304

Souza SM, Araújo OQF, Coelho MAZ (2008) Model-based optimization of a sequencing batch reactor for biological nitrogen removal. Biores Technol 99:3213–3223

Stone KC, Hunt PG, Szogi AA, Humenik EJ, Rice JM (2002) Constructed wetland design and performance for swine lagoon wastewater treatment. Trans Am Soc Agric Eng 45:723–730

Stone KC, Poach ME, Hunt PG, Reddy GB (2004) Marsh-pond-marsh constructed wetland design analysis for swine lagoon wastewater treatment. Ecol Eng 23:127–133

Ström L, Christensen TR (2007) Below ground carbon turnover and greenhouse gas exchanges in a sub-arctic wetland. Soil Biol Biochem 39:1689–1698

Sugai SF, Lindstrom JE, Braddock JF (1997) Environmental influences on the microbial degradation of Exxon Valdez oil on the shorelines of Prince William Sound, Alaska. Environ Sci Technol 31:1564–1572

Sun G, Zhao Y, Allen S, Cooper D (2006) Generating "tide" in pilot-scale constructed wetlands to enhance agricultural wastewater treatment. Eng Life Sci 6:560–565

Sundberg C, Tonderski K, Lindgren P-E (2007) Potential nitrification and denitrification and the corresponding composition of the bacterial communities in a compact constructed wetland treating landfill leachates. Wat Sci Technol 56:159–166

Szogi AA, Hunt PG (2001) Distribution of ammonium-N in the water-soil interface of a surface-flow constructed wetland for swine wastewater treatment. Wat Sci Technol 44:157–162

Szogi AA, Hunt PG, Humenik EJ (2000) Treatment of swine wastewater using a saturated-soil-culture soybean and flooded rice system. Trans Am Soc Agric Eng 43:327–335

Szogi AA, Hunt PG, Humenik FJ (2003) Nitrogen distribution in soils of constructed wetlands treating lagoon wastewater. Soil Sci Soc Am J 67:1943–1951

Szogi AA, Hunt PG, Sadler EJ, Evans DE (2004) Characteristics of oxidation-reduction processes in constructed wetlands for swine wastewater treatment. Appl Eng Agric 20:189–200

Tanner CC (1996) Plants for constructed wetland treatment systems – a comparison of the growth and nutrient uptake of eight emergent species. Ecol Eng 7:59–58

Tanner CC, Clayton JS, Upsdell MP (1995) Effect of loading rate and planting on treatment of dairy farm wastewaters in constructed wetlands: II. Removal of nitrogen and phosphorus. Wat Res 29:27–34

Tao W, Hall KJ, Masbough A, Frankowski K, Duff SJB (2005) Characterization of leachate from a woodwaste pile. Wat Qual Res J Can 40:476–483

Tao W, Hall KJ, Duff SJB (2006) Treatment of woodwaste leachate in surface flow mesocosm wetlands. Wat Qual Res J Can 41:325–332

Tao W, Hall K, Hall E (2007) Laboratory study on potential mechanisms for treatment of wood-waste leachate in surface flow constructed wetlands. J Environ Eng Sci 6:85–94

Taylor BR, Carmichael NB (2003) Toxicity and chemistry of aspen wood leachate to aquatic life: field study. Environ Toxicol Chem 22:2048–2056

Taylor BR, Goudey JS, Carmichael NB (1996) Toxicity of aspen wood leachate to aquatic life: laboratory studies. Environ Toxicol Chem 15:150–159

Thomas PR, Glover P, Kalaroopan T (1995) An evaluation of pollutant removal from secondary treated sewage effluent using a constructed wetland system. Wat Sci Technol 32:87–93

Thorén A-K, Legrand C, Tonderski KS (2004) Temporal export of nitrogen from a constructed wetland: influence of hydrology and senescing submerged plants. Ecol Eng 23:233–249

Throbäck NI, Enwall K, Jarvis Á, Hallin S (2004) Reassessing PCR primers targeting *nirS, nirK* and *nosZ* genes for community surveys of denitrifying bacteria with DGGE. FEMS Microbiol Ecol 49:401–417

Tokida T, Miyazaki T, Mizoguchi M, Seki K (2005) *In situ* accumulation of methane bubbles in a natural wetland soil. Eur J Soil Sci 56:389–395

Trias M, Hu Z, Mortula MM, Gordon RJ, Gagnon GA (2004) Impact of seasonal variation on treatment of swine wastewater. Environ Technol 25:775–781

Tunney H, Breeuwsma A, Withers PJA, Ehlert PAI (1997) Phosphorus fertiliser strategies: past, present and future. In: Tunney H, Carton OT, Brookes PC, Johnson AE (eds) Phosphorus losses from soil to water. CAB International, New York, pp. 358–361

Uğurlu M, Karaoğlu MH, Kula (2006) Experimental investigation of chemical oxygen demand, lignin and phenol removal from paper mill effluents using three-phase three-dimensional electrode reactor. Pol J Environ Stud 15:647–654

UNEP (2003) Convention on biological diversity. In: 9th Meeting of the Subsidiary Body on Scientific, Technical and Technological Advice (10–14 November 2003), Montreal, United Nations Environment Programme (UNEP)/CBD/SBSTTA/9/INF/4, New York, USA

USEPA (1995a) Constructed wetlands for wastewater treatment and wildlife habitat: 17 case studies. United States of America Environmental Protection Agency (USEPA) Number: 832R93005, http://www.epa.gov/owow/wetlands/pdf/hand.pdf. Accessed 1 Jan 2010

USEPA (1995b) Constructed wetlands for animal waste treatment: manual on performance, design, and operation with case histories. United States of America Environmental Protection Agency (USEPA) Number: 855B97001, http://www.epa.gov/owow/wetlands/pdf/hand.pdf. Accessed 1 Jan 2010

USEPA (1988) Constructed Wetlands and Aquatic Plant Systems for Municipal Wastewater Treatment. EPA/625/1-88/022. United States Environmental Protection Agency (USEPA), Office of Research and Development, Cincinnati, OH, USA

USEPA (2000). Constructed Wetlands Treatment of Municipal Wastewaters. EPA/625/R-99/010. United States Environmental Protection Agency (USEPA), Office of Research and Development, Cincinnati, OH, USA

USDA (1991) A Guide to Creating Wetlands for Agricultural Wastewater, Domestic Wastewater, Coal Mine Drainage Stormwater. Volume 1, General Considerations. United States Department of Agriculture. http://www.epa.gov/OWOW/wetlands/pdf/hand.pdf. Accessed 1 Jul 2009

Vacca G, Wand H, Nikolausz M, Kuschk P, Kastner M (2005) Effect of plants and filter materials on bacteria removal in pilot-scale constructed wetlands. Wat Res 39:1361–1373

Vrhovsek D, Kukanja V, Bulc T (1996) Constructed wetland (CW) for industrial wastewater treatment. Wat Res 30:2287–2292

Vymazal J (2007) Removal of nutrients in various types of constructed wetlands. Sci Total Environ 380:48–65

Wallace SD, Knight RL (2006) Small-scale constructed treatment systems: feasibility, design criteria, and O&M requirements. Final report, Project 01-CTS-5. Water Environment Research Foundation, Alexandria, VA, USA

Wang H, Gielen GJ, Judd ML, Stuthridge TR, Blackwell BG, Tomer MD, Pearce SH (1999) Treatment efficiency of land application for thermomechanical pulp mill effluent constituents. Appita J 52:383–386

WDOE (1995) Best management practices to prevent storm water pollution at log yards. Publication 95-53, Washington State Department of Ecology (WDOE), Olympia, WA, USA

Webber J, Gibbs J (1996) Water storage of timber: experience in Britain. Bulletin 117, Forestry Commission, Her Majesty's Stationery Office, London, UK

Weishampel P, Kolka R, King JY (2009) Carbon pools and productivity in a 1-km^2 heterogeneous forest and peatland mosaic in Minnesota, USA. For Ecol Manag 25:747–754

Welling I, Mielo T, Raisanen J, Hyvarinen M, Liukkonen T, Nurkka T, Lonka P, Rosenberg C, Peltonen Y, Svedberg U, Jappinen P (2001) Characterization and control of terpene emissions in Finnish sawmills. Am Ind Hyg Assoc J 62:172–175

Wemple C, Hendricks L (2000) Documenting the recovery of hydrocarbon-impacted wetlands: a multi-disciplinary approach. In: Means JL, Hinchee RE (eds) Wetlands and remediation: an international conference. Battelle, Columbus, OH, USA, pp. 73–78

Werker AG, Doughtery JM, Mchenry JL, Van Loon VA (2002) Treatment variability for wetland wastewater treatment design in cold climates. Ecol Eng 19:1–11

Wheeler EF, Topper PA, Graves RE, Bruns MA, Wysocki CJ (2007) Odour-reduction performance of constructed wetlands treating diluted swine manure. Appl Eng Agric 23:621–630

Woodhouse C, Duff JBS (2004) Treatment of log yard runoff in an aerobic trickling filter. Wat Qual Res J Can 39:230–236

Wu Y, Chung A, Tama NFY, Pia N, Wong MH (2008) Constructed mangrove wetland as secondary treatment system for municipal wastewater. Ecol Eng 34:137–146

Xia H (2003) Ecological effectiveness of constructed wetlands in treating oil-refined wastewater. In: Proceedings of the 3rd international conference in Guangzhou, China, 6–9 Oct 2003

Yang L, Chang H, Huang ML (2001) Nutrient removal in gravel- and soil-based wetland microcosms with and without vegetation. Ecol Eng 18:91–105

Zedler JB (2003) Wetlands at your service: reducing impacts of agriculture at the watershed scale. Front Ecol Env 1:65–72

Zenaitis MG, Duff SJB (2002) Ozone for removal of acute toxicity from log yard run-off. Ozone Sci Eng 24:83–90

Zenaitis MG, Sandhu H, Duff SJB (2002) Combined biological and ozone treatment of log yard runoff. Wat Res 36:2053–2061

Zelikoff JT, Chen LC, Cohen MD, Schlesinger RB (2002) The toxicology of inhaled woodsmoke. J Toxicol Environ Health B Crit Rev 5:269–282

Zhang L, Scholz M, Mustafa A, Harrington R (2009) Application of the self-organizing map as a prediction tool for an integrated constructed wetland agroecosystem treating agricultural runoff. Biores Technol 100:539–565

Zhao YQ, Sun G, Allen SJ (2004) Anti-sized reed bed system for animal wastewater treatment: a comparative study. Wat Res 38:2907–2917

Zheng J, Nanbakhsh H, Scholz M (2006) Case study: design and operation of sustainable urban infiltration ponds treating storm runoff. J Urban Plann Develop Am Soc Civ Eng 132:36–41

Zhou E, Crawford RL (1995) Effects of oxygen, nitrogen, and temperature on gasoline biodegradation in soil. Biodegradation 6:127–140

Chapter 3
Carbon Storage and Fluxes
Within Wetland Systems

Abstract This chapter critically reviews recent literature on carbon storage and fluxes within natural and constructed freshwater wetlands and specifically addresses concerns of readers working in the field of applied science. The purpose is to review and assess the distribution and conversion of carbon in the water environment, particularly within constructed wetland systems. A key aim is to assess if wetlands are carbon sinks or sources. Carbon sequestration and fluxes in natural and constructed wetlands located around the world are assessed. All facets of carbon (solid and gaseous forms) have been covered. Conclusions are based on these studies. Findings indicate that wetlands can be both sources and sinks of carbon, depending on their age, operation, and the environmental boundary conditions such as location and climate. Suggestions for further research needs in the area of carbon storage in wetland sediments are outlined to facilitate the understanding of the processes of carbon storage and removal and also the factors that influence them. This timely chapter should help engineers to make the right decisions when designing wetlands taking climate change into consideration.

3.1 Introduction

3.1.1 Wetlands and Processes

Wetlands are areas of water-saturated soil and include small lakes, floodplains, and marshes. Wetlands only cover a small proportion of the Earth's land surface (between approx. 2 and 6%, depending on definitions) but contain a large proportion of the world's carbon (approx. 15×10^{14} kg) stored in terrestrial soil reservoirs (Schlesinger 1991; Amthor *et al.* 1998; Whiting and Chanton 2001). Wetlands play an important role in carbon cycling because they represent 15% of the terrestrial organic matter losses to the oceans (Hedges *et al.* 1997; Stern *et al.* 2007). Among all terrestrial ecosystems, they have the highest carbon density. Further-

M. Scholz, *Wetland Systems*
© Springer 2011

more, wetlands are a diffuse source of humic substances for some receiving freshwater systems (Stern *et al.* 2007).

Decomposition within wetlands is a complicated process as it involves aerobic and anaerobic processes. Organic matter decomposition is often incomplete under anaerobic conditions. The lack of oxygen is therefore the main factor determining plant detritus turnover. Consequently, plant remains coming from the inflow, the wetland biomass, or from the vegetation growing along the wetland margins accumulate within the wetland system, and different decomposition stages can be identified (Gorham *et al.* 1998; Collins and Kuehl 2001; Holden 2005). A net retention of organic matter and plant detritus can be observed in most wetlands (Mitsch and Gosselink 2007). Organic matter accumulation in wetland sediments depends on the ratio between inputs (organic matter produced *in situ* and *ex situ*) and outputs. The latter may be due to the decomposition under waterlogged conditions, erosion due to high precipitation, and soil disturbance in general (Gorham *et al.* 1998).

Since 1980, treatment wetland systems have gained popularity and have been applied successfully for the treatment of numerous waste streams (Kadlec *et al.* 2000; Haberl *et al.* 2003; Zhang *et al.* 2005; Vymazal 2007) and runoff from urban areas (Scholz 2006), farmyards (Carty *et al.* 2008), and log yards (Hedmark and Scholz 2008). The concept of constructed wetlands applied for the purification of wastewaters has received growing interest because most of these systems are easy to use, require only little maintenance, and have low construction costs (Machate *et al.* 1997). Dissolved organic matter is a very important water quality parameter associated with the performance of treatment wetland systems. Some microorganisms including bacteria use dissolved organic matter as an energy source for processes such as denitrification. However, too high levels of dissolved organic matter can prevent light penetration within the water column (Pinney *et al.* 2000; Li *et al.* 2008). The treatment efficiencies of wetlands vary depending on climate, vegetation, microorganism communities, and type of wetland system (Waddington *et al.* 1996; Schlesinger 1997; Joabsson *et al.* 1999; Trettin and Jurgensen 2003; Whalen 2005; Picek *et al.* 2007; Ström and Christensen 2007; Weishampel *et al.* 2009).

Scientists have carried out detailed investigations concerning wetland biochemistry and hydrology. Nevertheless, there is no commonly accepted agreement if wetlands are actually carbon sources or sinks. There is disagreement in the interpretation of variables, reactions, and the impact of environmental conditions on carbon storage and release. Therefore, recommendations on how to adapt policies and planning processes to enhance carbon storage vary considerably. Comparisons of carbon storage and flux data vary greatly as a function of region and climate (Kayranli *et al.* 2010).

3.1.2 Global Warming

Global warming mitigation is becoming increasingly important as the effects of climate change are becoming apparent around the world. Depending predominantly on the meteorological and hydrological conditions, wetlands can absorb carbon dioxide from the atmosphere and capture it within the sediment, and may therefore be greenhouse gas sinks. The high productivity, high water table, and low decomposition rate associated with wetlands lead to carbon storage within the soil, sediment, and detritus (Whiting and Chanton 2001). The process of locking carbon dioxide away from the atmosphere is called carbon sequestration (Kayranli *et al.* 2010).

On the other hand, wetlands are considered to be greenhouse gas sources particularly with respect to the emission of methane gas to the atmosphere. Methane has a much higher global warming potential than carbon dioxide and contributes to the atmospheric sorption of infrared radiation and subsequent warming (Carroll and Crill 1997; Whiting and Chanton 2001; Zhang *et al.* 2005). Minimizing methane fluxes from created and restored wetlands should therefore be a vital aim in combating climate change. Improved design, construction, and operation of wetlands used for treatment and conservation purposes should therefore help to mitigate global warming by reducing the release of greenhouse gases and enhancing carbon storage at the same time (Kayranli *et al.* 2010).

3.1.3 Purpose and Review Methodology

This review focuses on the assessment of the key processes determining carbon removal, sequestration, and fluxes within wetlands. It has specifically been written for applied scientists and engineers working worldwide and complements other more ecology-based papers such as that by Bridgham *et al.* (2006) focusing on Northern America. The aims of the key sections are:

- to discuss carbon turnover and removal processes within wetlands;
- to highlight processes where wetlands can be described as carbon sources or carbon sinks;
- to discuss the effect of global warming on wetlands.

3.2 Carbon Turnover and Removal Mechanisms

3.2.1 Carbon Turnover

Various reactions utilizing carbon take place within wetlands. The key processes are respiration in the aerobic zone, fermentation, methanogenesis, and sulfate, iron, and nitrate reduction in the anaerobic zone. Organic matter typically contains between 45 and 50% carbon. Wetlands contain large amounts of dissolved organic matter, promoting microbial activity (Bano *et al.* 1997; Zweifel 1999). Bacterial oxidation of dissolved organic carbon subsequently results in mineralization, which is a process whereby organic substances are converted into inorganic substances (Hensel *et al.* 1999).

Respiration is the biological conversion of carbohydrates into carbon dioxide, and fermentation is the conversion of carbohydrates into chemical compounds such as lactic acid, or ethanol and carbon dioxide. In a wetland, organic carbon is converted into compounds including carbon dioxide and methane or stored in plants, dead plant matter, microorganisms, or peat. A significant part of the BOD may be particle-bound and, therefore, susceptible to removal by particulate settling (Kadlec *et al.* 2000; Kayranli *et al.* 2010).

3.2.2 Carbon Components

Wetlands contain five main carbon reservoirs: plant biomass carbon, particulate organic carbon, dissolved organic carbon, microbial biomass carbon, and gaseous end products such as carbon dioxide and methane. The latter four are present in water, detritus, and soil (Kadlec and Knight 1996). Wynn and Liehr (2001) outlined a carbon cycle comprising the following key components: plant biomass, standing dead plants, particulate organic carbon, dissolved organic carbon, and refractory carbon (*i.e.*, resistant carbon, which would retain its strength at high temperatures). These carbon reservoirs can be used in the description of carbon cycles (Kayranli *et al.* 2010).

Active biomass may comprise wetland plants and periphyton (microorganisms and detritus attached to submerged surfaces) and contributes to the transformation of inorganic carbon such as carbon dioxide into organic carbon through photosynthesis. The productivity of wetlands varies due to the time of year, geographic location, nutrient status, and type of vegetation. Particulate organic carbon consists of decaying plant matter, microbial cells, particulate influent, and particulate organic substances found on the soil surface. Dissolved organic carbon comprises dissolved BOD and other carbon components in solution. While dissolved organic carbon typically represents <1% of the total organic carbon in soil, it represents approx. 90% of the total organic carbon in surface waters (Kadlec and Knight 1996; Wynn and Liehr 2001; Reddy and Delaune 2008). Microbial biomass carbon occurs in heterotrophic microfloral catabolic activities, transforming organic carbon (energy reserve of the ecosystem) back into inorganic carbon and mineralizing particulate organic carbon and dissolved organic carbon (D'Angelo and Reddy 1999; Picek *et al.* 2007). The turnover of active biomass happens relatively quickly, usually in the order of days, while the corresponding turnover of soil organic matter takes decades. Soil microbial biomass can be regarded as a significant carbon sink (Kayranli *et al.* 2010).

3.2.3 Carbon Removal Mechanisms

Carbon processing in the wetland environment is complex, and the various decomposition reactions take place in different horizons; *e.g.*, respiration and methane oxidation occur in the aerobic zones while methanogenesis occurs in anaerobic zones (Knight and Wallace 2008). However, the highest rates of decomposition are found closest to the wetland surface where there is an elevated input of fresh litter and recently synthesized labile organic matter (Sherry *et al.* 1998).

The organic matter content within wetland systems is impacted by processes such as biodegradation, photochemical oxidation, sedimentation, volatilization, and sorption. Some of these mechanisms provide natural organic matter accumulation via microbial or vegetative decay (Burgoon *et al.* 1995; Reddy and D'Angelo 1997; Stottmeister *et al.* 2003; Quanrud *et al.* 2004; Li *et al.* 2008). Moreover, the accumulation of organic matter is a potential energy source for microbial communities (Turcq *et al.* 2002; Reddy and Delaune 2008).

Dissolved organic matter degradation is expected to occur via heterotrophic uptake by aerobic and anaerobic bacteria, and degradation by ultraviolet light. Several authors have reported on dissolved organic matter transformations in algae (Kragh and Søndergaard 2004), forest vegetation (Li *et al.* 2008), *Typha* spp. wetland plant material (Pinney *et al.* 2000), microbial groups (Ibekwe *et al.* 2003; Li *et al.* 2008), and soils (Qualls and Haines 1992). Dissolved organic matter from plant exudates appears more dominant during warm months with active plant growth (Pinney *et al.* 2000).

Organic matter accumulates when primary productivity is faster than the corresponding decomposition rate, leading to a net accumulation of organic matter (Mitsch and Gosselink 2007). Due to slow organic mater decomposition rates, strata are built up and compressed to form different soil layers. Organic matter from inflow and wetland plants is accumulated, decomposed, and subsequently buried in the system. This results in a shift from aerobic to anaerobic processes due to lack of oxygen in the wetland sediment, which drastically reduces decomposition rates (Holden 2005). It is also believed that some parameters such as temperature, organic matter quality, residence time of organic matter in the water column, vegetation pattern, wetland maturity, sedimentation rate, sediment texture, and sediment reworking impact the organic matter decomposition within the water body and the organic matter compositions (Borman *et al.* 1995; van der Peijl and Verhoeven 1999; Barber *et al.* 2001; Savage and Davidson 2001; Turcq *et al.* 2002; Yu *et al.* 2002; Lafleur *et al.* 2005; Wolf and Wagner 2005; Shepherd *et al.* 2007; Yurova and Lankreijer 2007). Furthermore, organic matter compositions consist of labile and resistant fractions within the soil profile. Many of the labile compounds are accumulated on the sediment surface and decomposed within a few months (Schlesinger 1997; Wolf and Wagner 2005).

Wetlands have aerobic and anaerobic interfaces in water, soil, and the accumulated organic matter (Scholz *et al.* 2007). Gaseous end products are formed under anaerobic and aerobic conditions. Under anaerobic conditions, carbon dioxide and methane are formed through the decomposition of organic matter. In comparison, under aerobic conditions, only carbon dioxide is formed. Previous researchers (Kadlec and Knight 1996; Scholz 2006; Mitsch and Gosselink 2007) pointed out that the aerobic respiration in wetland systems is far more effective with respect to organic matter degradation than anaerobic processes such as fermentation and methanogenesis.

The dissolved organic carbon cycle depends on the cycling and bioavailability of phosphorus and nitrogen (Craft and Richardson 1998) as well as on the bioavailability and transport of metals (Voelker and Kogut 2001; Tipping and Center 2002). Microbial death is generally assumed to only contribute to particulate organic matter and not to dissolved organic matter because most bacteria in wetlands are associated with plant litter and soil organic matter. When the growing season reaches its end, approx. 15% of the plant carbon disappears due to leaching and physical degradation in temperate climates (Kadlec and Knight 1996). The remainder degrades over approx. 1 year and becomes predominantly particulate carbon (Johnston 1991; Wynn and Liehr 2001).

3.3 Are Wetlands Carbon Sources or Sinks?

3.3.1 Wetlands as Carbon Sources

The important greenhouse gases carbon dioxide, methane, and nitrous oxide can be released from natural and constructed wetlands (Le Mer and Roger 2001; Whiting and Chanton 2001; Malmer et al. 2005; Liikanen et al. 2006; Mander et al. 2008). Processes such as denitrification and methane production are dependent on the oxygen status of soil and sediment. Anoxic soils and sediments produce methane, while well-drained soils act as a sink for atmospheric methane due to methane oxidation (Hanson and Hanson 1996).

The water table level of wetlands influences not only the amount of methane emitted to the atmosphere but also the removal of methane from the atmosphere. For example, Harris et al. (1982) determined that peat from the Great Dismal Swamp contributes to the removal of atmospheric methane when the water table level is below the surface of peat during dry periods. In contrast, when peat is well saturated with water, it becomes an important methane source. Furthermore, Augustin et al. (1998) concluded that lowering the groundwater table of minerotrophic fens in the northeast of Germany increased the release of nitrous oxide and the reduction of methane. Relatively high methane emissions could be observed when the groundwater table was high and soil temperatures were higher than 12°C. They also point out that these fens release approx. between 0.6 and 9.0 mg CH_4–$C/m^2/h$.

Natural wetlands emit approx. 1.45×10^{11} kg CH_4–C/a to the atmosphere. This equates to about 25% of the total emissions from all anthropogenic and natural sources. Wetland methane flux rates are commonly 10^{-6} kg/m^2/d and represent the net effects of microbial production and consumption (Whalen 2005).

The studies carried out so far have demonstrated that constructed and restored wetlands also have high nitrous oxide (Fey et al. 1999; Xue et al. 1999; Johansson et al. 2003; Mander et al. 2005; Stadmark and Leonardson 2005; Teiter and Mander 2005; Liikanen *et al.* 2006; Picek *et al.* 2007) and methane (Schipper and Reddy 1994; Tanner *et al.* 1997; Cao *et al.* 1998; Tuittila *et al.* 2000; Johansson *et al.* 2004; Mander *et al.* 2005; Teiter and Mander 2005; Altor and Mitsch 2006; Liikanen *et al.* 2006; Picek *et al.* 2007) emissions. However, data are highly variable due to different designs and operations, and more importantly as a result of system locations in different climates. The methane emissions from vegetated constructed treatment wetlands can be similar to those from productive natural wetlands. Methanogenesis can be an important decomposition process in constructed wetlands treating organic wastewaters. For example, a flux of approx. between 28 and 278 mg CH_4–$C/m^2/h$ has been reported for pilot-scale constructed wetlands (Tanner *et al.* 1997).

Brix *et al.* (2001) reported that in wetlands dominated by *Phragmites australis*, the primary productivity is high, and approx. 50% of the net primary production is respired to carbon dioxide and methane in the sediment. In the growing season, the process of methanogenesis is primarily limited by organic matter availability, while at other times temperature is the most important factor.

Like natural wetlands, rice paddies have been identified as one of the important sources of atmospheric methane. The majority of studies reporting on methane emissions from wetlands have been conducted in natural ecosystems or heavily managed rice paddies (Cao *et al.* 1998; Crutzen 1995). Cao *et al.* (1998) estimated that the global annual methane emission from wetlands is 1.45×10^{11} kg, of which 0.92×10^{11} kg comes from natural wetlands and 0.53×10^{11} kg from rice paddies.

A limited number of studies consider carbon dioxide fluxes from constructed wetlands (Mander *et al.* 2005; Liikanen *et al.* 2006). Many authors report that carbon dioxide emissions increase with increasing temperature and are higher under drained than flooded conditions (Bridgham and Richardson 1992; Moore and Dalva 1993, 1997; Price and Waddington 2000; Scanlon and Moore 2000; Waddington *et al.* 2001).

Deep wetlands generally capture carbon dioxide from and release methane into the atmosphere (Whiting and Chanton 2001). The combination of these two fluxes determines whether these countervailing processes make a wetland system an overall contributor to the greenhouse effect. The ratio of methane release to carbon dioxide consumption determines the carbon exchange balance with the atmosphere for any wetland ecosystem (Kayranli *et al.* 2010).

A better understanding of the critical processes regulating greenhouse gases associated with wetlands such as freezing–thawing cycles and pulsing hydrological regimes are important for assessing carbon dioxide, methane, and nitrous oxide fluxes. The production and consumption of greenhouse gases are partly regulated by microbial processes, which in turn are influenced by soil moisture and temperature. Nitrification and denitrification are the key processes that produce nitrous oxide. However, nitrous oxide production in frozen soils is more likely to be regulated by denitrification (Mørkved *et al.* 2006; Öquist *et al.* 2007). Van Bochove *et al.* (2001) highlighted that nitrous oxide fluxes are high in winter because of the sudden release of stored nitrous oxide. Maljanen *et al.* (2007) reported that nitrous oxide and carbon dioxide accumulated in the soil during winter and were released swiftly during thawing in spring. During winter, methane concentrations in the soil remained lower than in the atmosphere and subsequently increased as temperatures increased after thawing.

Zhang et al. (2005) observed that during thawing, methane and carbon dioxide emissions increased rapidly (4.5 to 6 times the winter emissions) for continuously flooded and seasonally flooded marshes. They estimated that a continuously flooded and a seasonally flooded wetland in Sanjiang (northeast China) released 0.5 ± 0.19 and 0.18 ± 0.15 mg CH_4–C/m^2/h methane, respectively. In comparison, naturally flooded forests and floating grass mats in Brazil (Amazon floodplain) emitted between 8 and 92 mg CH_4–C/m^2/h into the atmosphere (Bartlett et al. 1988).

Environmental parameters such as temperature, pH, depth of water table, planting regime (Waddington et al. 1996; Schlesinger 1997; Trettin and Jurgensen 2003; Whalen 2005; Inamori et al. 2007; Picek et al. 2007; Knoblauch et al. 2008), substrate type and quality (Bellisario et al. 1999; Joabsson et al. 1999; Ström et al. 2003), and specialized microbes (Fischer and Pusch 1999; Whalen 2005; Buesing and Gessner 2006; Picek et al. 2007; Sleytr et al. 2007; Ström and Christensen 2007; Tietz et al. 2008) impact on gas production and, ultimately, net methane emission rates. Furthermore, the methane exchange between wetland ecosystems and the atmosphere can be affected by the presence of plants because the convective flow process in plants facilitates a faster diffusion of gases through water, and particularly by the species composition of vascular plants. These plants affect important aspects of methane dynamics such as production, consumption, and transport; for example, the root exudates are decomposed by microbes and transformed into methane and carbon dioxide (Zhu and Sikora 1995; Joabsson and Christensen 2001; Tanner 2001; Picek et al. 2007; Ström and Christensen 2007).

Walter and Heimann (2000) emphasize that most wetland plants root below the water table and that methane flux from the soil to the atmosphere takes place via aerenchyma in the vascular tissue of the plants. Additional transport pathways are diffusion and bubble generation. Above the water table, methane is oxidized in the soil pores by methanotrophic bacteria.

Landry et al. (2009) claim that constructed wetlands emit between 2 and 10 times more greenhouse gases than natural wetlands. This is likely due to high loading rates. They observed that methane was the most important greenhouse gas in unplanted wetland systems and that the presence of plants decreased methane fluxes but favored carbon dioxide production.

Alford et al. (1997) estimated fluxes of between approx. 6 and 38 mg CH4–C/m^2/h for swamp forests and marshes near New Orleans, LA, USA. Barlett and Harris (1993) reported fluxes of approx. 4 mg CH4–C/m^2/h for forested swamps and marshes. Kang and Freeman (2002) reported that bog and forested swamps in North Wales (UK) emit up to approx. 3 mg CH4–C/m^2/h into the atmosphere. The relatively high data variability is likely due to different climatic regions.

Hou et al. (2000) pointed out that the reduction of various oxidants in homogeneous soil suspensions occurs sequentially at corresponding soil redox potentials. The availability of soil oxidants such as oxygen and carbon dioxide used as electron acceptors for organic matter degradation contributes significantly to soil microbiological processes. They found that emissions of methane were strongly correlated with changes in the soil redox potential. Significant methane emissions occurred only at soil redox potentials, which were lower than approx. -100 mV.

It is the water table level that largely determines the presence of aerobic and anaerobic conditions occurring at different depths of wetlands. These conditions control the methanogenic and methanotrophic processes (Kelley *et al.* 1995). Methanogenesis is a rigid anaerobic process and is evoked during flooding periods, when the water table level rises. In contrast, with a decrease in flooding periods, methane production decreases. An inverse relation is observed for methane oxidation (Kayranli *et al.* 2010).

Grünfeld and Brix (1999) compared vegetated organic sediments at different water table depths below the surface with vegetated inundated sediments. They found that due to the high water-holding capacity of organic sediments, rates of methanogenesis and methane emission in organic sediments with a water table of 8 cm below the sediment surface were only slightly, but not statistically significantly, different from rates in inundated sediments. Sandy sediments with water tables of 8 cm below the sediment surface had very low methanogenic activity as compared with organic sediments.

Methane can be transported to the atmosphere through pathways such as molecular diffusion, gas bubbling up (ebullition) from the sediments, and vascular plant stems (Walter and Heimann 2000). King (1996) pointed out that the amount of methane oxidized did not correlate with the total potential methane fluxes from a wetland. Oxygen distribution and availability controls the rates of methane oxidation within wetlands. Moreover, oxygen penetration within peat varies from 1 to 7 mm, with some diurnal variation coupled to benthic photosynthesis.

Moore and Dalva (1993) and Moore and Roulet (1993) reported that the mean position of the water table level is the best indicator of methane emissions. Apparently, a critical depth exists at which maximal emissions occur. It has been determined that a water table depth greater than 18 cm does not produce high emissions, since methane production (methanogenesis) decreases and its consumption increases (methanotrophy). However, when the depth of the water table is 12 cm below the surface of peat, or exceeds it, methane fluxes are high. Bubier *et al.* (1993) and Daulat and Clymo (1998) estimated that methane emitted into the atmosphere from experimental digs was between 5 and 60 times higher than that produced in hillocks (small hills or mounds) due to the digs' having a lower water table depth than the hillocks. Roulet *et al.* (1993) discovered that peatlands are converted from a source into a sink of methane when the water table drops to 25 cm below the peat surface due to increased methane oxidation. Kelley *et al.* (1995) studied methane emissions across a tidally flooded riverbank in North Carolina, USA. Their study showed the highest methane fluxes when the water level was close to the surface and the lowest fluxes at both high and low water table levels. Similarly, Smith *et al.* (2000) estimated that methane emissions stopped when the soil moisture content fell below approx. 25%, as floodwaters receded in Venezuela's Orinoco River floodplain.

Glatze *et al.* (2004) pointed out that the highest rates of anaerobic methane production can be measured for samples close to the soil surface with fresh peat accumulation and a high water table. In contrast, the lowest rates were observed for samples from the sub-surface of sites with a low water table. Anaerobic methane production was significantly positively correlated with aerobic and anaerobic carbon dioxide production. These production potentials show that drainage (Salm *et al.* 2009), harvesting, and restoration change the ability of the peat profile to produce and emit carbon dioxide and methane.

3.3.2   Wetlands as Carbon Sinks

The results indicate that wetlands are vital carbon sinks. The majority of studies on carbon sequestration within wetlands focus on sediment, soil, and living plant communities (Krogh *et al.* 2003; Brevik and Homburg 2004; Bedard-Haughn *et al.* 2006; Euliss *et al.* 2006; Alongi *et al.* 2007).

Freeman *et al.* (2001a, 2004) pointed out that phenol oxidase is the main enzyme that by remaining inactive, keeps much of the world's terrestrial wetland carbon locked up. Actinomyces (filamentous, mostly anaerobic microorganisms of this genus), bacteria, and certain fungi are direct indicators of decomposer activity; they excrete extracellular enzymes to decompose complex high-molecular-weight compounds. The activity of phenol oxidase increases with increasing temperatures (Freeman *et al.* 2001b) and is reduced by low pH (Pind *et al.* 1994; Williams *et al.* 2000). Williams *et al.* (2000) studied phenol oxidase activity in *Sphagnum* spp. peat and reported that when pH was favorable, the activity of phenol oxidase depended more on the botanical composition of the peat and the wetland vegetation type than on the water level. Wetland carbon stores (especially peatlands) may become considerable methane sources when aerobic soil conditions activate the phenol oxidase enzyme, which triggers chain reactions breaking down lignin and humic substances, releasing methane into the atmosphere in substantial amounts (Freeman *et al.* 2004). Some models of small-scale constructed wetlands show that they sequester very small amounts of carbon. However, they are considered sizeable carbon sinks due to the difference in energy consumption between the wetland and the equivalent wastewater treatment plant. Thus, small-scale constructed wetlands used for the treatment of wastewater are considered carbon sinks (Ogden 2001). The main factors controlling methane emissions from wetlands are soil temperature (Christensen *et al.* 2003), water table depth (Moore *et al.* 1998), and the amount and quality of decomposable substrate (Christensen *et al.* 2003). The factors controlling methane oxidation are well documented by Boon and Lee (1997); they include the supply of oxygen and temperature. Methane oxidation rates can be optimized by promoting well-aerated water columns and, in turn, well-aerated sediments. Furthermore, nitrate affects oxidation, but only at relatively high nutrient concentrations, and the availability of ammonium and sulfate has little or no effect on oxidation rates (Kayranli *et al.* 2010). Hanson and Hanson (1996) documented that wetland soils are normally fully saturated and are often located well below the water table. These wetland conditions create mainly anaerobic or anoxic soils, which store carbon dioxide and release methane. However, drained wetlands, which have unsaturated soils, are atmospheric methane sinks. Methane is absorbed through methanotrophs and anaerobic methane-oxidizing bacteria (Kayranli *et al.* 2010). Some studies identified variable methane fluxes from wetlands in Canada. For peatlands, bogs, and fens, 2.8 ± 0.27 mg CH_4–$C/m^2/h$ was calculated by Turetsky *et al.* (2002), for bogs and rich fens, between 1 and 10 mg CH_4–$C/m^2/h$ was estimated by Bellisario *et al.* (1999), for fens, bogs, ponds, and palsa (*i.e.*, low and oval rise occurring in polar climates), releases up to 11 mg CH_4–$C/m^2/h$ were published by Liblik *et al.* (1997), and peatlands released up to 15 mg CH_4–$C/m^2/h$ according to Moore and Roulet (1995). However, for freshwater wetlands, a flux of only approx. 0.3 mg CH_4–$C/m^2/h$ was estimated by Bridgham *et al.* (2006). Concerning wetlands in the USA, Armentano and Menges (1986) estimated fluxes of roughly 5 mg CH_4–$C/m^2/h$ for northern peatlands in northern territories and releases of roughly 26 mg CH_4–$C/m^2/h$ for Florida. Again high data variability reflects different wetland types located in various climatic regions. On the other hand, Freeman *et al.* (2004) pointed out that when the water table drops considerably below the peatland surface, peatlands may change from being a source to a sink for methane due to increased methane oxidation. However, drought conditions in a peatland lower methane emissions during the drawdown of the water table due to decreased methanogenesis rather than methane consumption (Kayranli *et al.* 2010). Flooded wetlands generally sequester carbon dioxide from and release methane into the atmosphere. The combination of these two factors determines whether these offsetting processes make a wetland system an overall contributor to the greenhouse effect. Maximizing permanent vegetation in cultivated wetlands could provide maximum carbon sequestration, but the overall consequences for the gas emissions need to be carefully assessed (Bedard-Haughn *et al.* 2006). McCarty and Ritchie (2002) claimed that agricultural activity increased the rate of carbon storage within the sediment and contributed to the accumulation of nutrients within a wetland ecosystem. In comparison, Bedard-Haughn *et al.* (2006) concluded that organic carbon densities decreased from uncultivated to cultivated wetlands. McCarty and Ritchie (2002) also reported that an agricultural field and a riparian ecosystem in Maryland (USA) sequestered between 0.16 and 0.22 kg $C/m^2/a$. Moreover, prairie wetlands in the northcentral USA are known to have sequestration values of roughly 0.3 kg $C/m^2/a$ (Euliss *et al.* 2006). Wetlands store approx. twice the organic carbon load in comparison to cropland that is not tilled (Euliss *et al.* 2006). For example, northern peatlands in Scandinavia are important carbon stores. These peatlands often show large spatial and temporal variation in the atmospheric exchange of carbon dioxide and methane. The main parameters impacting carbon storage within these wetlands are erosion and soil movement (McCarty and Ritchie 2002), excessive drainage (Salm *et al.* 2009), water discharge, and nutrient input (Turcq *et al.* 2002).

Most of methane and carbon dioxide fluxes take place in the relatively thin oxic layers near the surface of peatlands. In the oxic surface layers of peatlands, the rates of litter decomposition may not generally differ from those found in the mineral soil sites for the same litter types (Moore *et al.* 2002; Vavrova *et al.* 2009). The large amounts of carbon captured within peatlands and their low productivity highlight the potential of peatlands to significantly impact regional carbon cycling, particularly at times when climate change might lead to increased peat degradation due to increased temperatures and lower water tables (Gorham 1991; Weishampel *et al.* 2009). Raghoebarsing *et al.* (2005) have shown that methane consumption by methanotrophic bacteria living in symbiosis with some *Sphagnum* species leads to effective *in situ* methane recycling within peatlands. These findings have also helped to explain the high organic carbon burial within wetland ecosystems. In a subsequent paper, Raghoebarsing *et al.* (2006) demonstrated that the direct anaerobic oxidation of methane coupled to denitrification of nitrate was possible. The reactions presented make a substantial contribution to the microbiological methane cycle. Landry *et al.* (2009) reported that planted wetlands may sequester between 2 and 15 times more carbon than they emit as carbon dioxide. However, respiration by stems and leaves, which was not accounted for in this study, could have reduced the reported carbon sequestration values. Moreover, they observed that methane was the most important greenhouse gas in unplanted wetland systems. They also found that the presence of plants decreased methane fluxes but favored carbon dioxide production. The carbon sequestration potential of swamps is usually much higher than that of lakes. The accumulation of carbon within lake sediments depends on the water table height and on the regional climate. While low carbon storage occurs in drier climates, humid climates bring about high carbon accumulation within most lakes (Turcq *et al.* 2002). Based on a wide range of assumptions, Mitra *et al.* (2005) calculated the net balance between methane production and carbon sequestration in the world's wetlands and deduced that the overall impact of wetlands on climate change in the carbon cycle was minimal. High numbers of spatially distinct samples for carbon sequestration (Anderson and Mitsch 2006) and methane generation (Altor and Mitsch 2006) were collected from two created wetlands (Ohio, USA), and subsequent calculations were based on conversions proposed by Mitra *et al.* (2005). It was found that the created wetlands were climate neutral or even had a cooling effect.

3.4 Impact of Global Warming on Wetlands

Wetland soils and sediments are considered to be among the world's largest carbon sinks. They have been accumulating carbon for between 4000 and 5000 years (Lloyd 2006), but are at risk of becoming an extremely large atmospheric carbon source because of climate change. Peatlands store an estimated one third of the world's organic soil carbon (Gorham 1991; Weishampel *et al.* 2009). Weishampel *et al.* (2009) stated that carbon storage in peatlands is a consequence of long-term climate trends during which a positive water balance enables accretion of peat. Although open peatland productivity is low, all peatlands have acted as long-term carbon sinks for hundreds to thousands of years and store significantly more carbon per unit area than is stored in uplands. Over long periods of time, natural wetlands can be considered carbon stores (Kayranli *et al.* 2010).

There is a strong correlation between climate and soil carbon pools, where the organic carbon content decreases with increasing temperatures (Kirschbaum 1995; Rasmussen *et al.* 1998) due to decomposition rates doubling with every 10°C increase in temperature (Schlesinger 1997; Hartel 2005). If temperature increases continue and become more rapid, the decomposition of organic matter will increase, and wetlands will eventually become major sources of carbon. Some researchers (Gorham 1991; Hobbie *et al.* 2000; Davidson and Janssens 2006) point out that wetlands, which drain well and are therefore well aerated, will be associated with fewer fluxes of carbon dioxide in the event of warming. However, if wetland drainage is poor and anaerobic conditions occur within the soil, wetlands may release considerable amounts of greenhouse gases. There is a general belief that increases in temperature and changes in water levels are important variables in the production of methane and carbon dioxide from wetlands (Moore and Roulet 1995; Updegraff *et al.* 2001). For example, as a result of a slight global temperature rise, parts of the tundra environment would act as a net source of carbon dioxide (Christensen 1993).

Trenberth *et al.* (2007) estimated that land warming in the Arctic is expected to be twice as high as the global mean, and thus the effects of the observed and predicted climate changes will be particularly strong in the Arctic. Methane emissions from Arctic wetlands are expected to increase (Wuebbles and Hayhoe 2002), and highly variable emissions, potentially indicating signs of climate change, have already been recorded for some sub-arctic wetlands. Ström and Christensen (2007) reported between 0.2 and 36.1 mg CH_4–$C/m^2/h$ for northern parts of Sweden. Dick and Gregorich (2004) carried out research to compare the decomposition rates of organic matter in tropical regions of Nigeria and cold dry climates in Canada, and concluded that decomposition rates were usually ten times faster in tropical regions than in cold and dry climates. Hence, global warming effects on tropical wetlands may also lead to increased decomposition and carbon fluxes, unless the corresponding temperature change is modest (Kayranli *et al.* 2010).

The impact of global warming on the economic exploitation of wetlands and on conservation policies is not well understood and is therefore often not considered in global models of climate-change effects (Clair *et al.* 1998). Much of recent terrestrial ecosystem modeling is aimed at estimating ecosystem carbon budgets and their future trends under a changing climate. Moreover, the current global financial crisis is likely to lead to reduced investment in wetland protection and conservation measures. The future of conservation wetlands should be secured by protecting their status (Kayranli *et al.* 2010).

The high initial global warming potential of increased methane emission in newly created wetlands means that many will have to establish themselves for over 100 years to be considered as carbon sinks (Whiting and Chanton 2001). However, improved design and management of constructed treatment wetlands, even after their decommissioning, should make a positive impact on long-term carbon storage. Changes in land use due to global warming such as increased drainage of wetlands for agricultural purposes could potentially lead to large carbon dioxide and methane fluxes to the atmosphere, further accelerating climate change (Limpens *et al.* 2008). Destruction of wetlands is also likely to lead to secondary water pollution from the release of nutrients during wetland degradation due to lower water levels, as demonstrated at a peatland restoration project in Northern Germany (Scholz and Trepel 2004a, b).

3.5 Conclusions and Further Research Needs

Different types of wetland systems such as natural, constructed, treatment, and integrated ones have the potential to sequester carbon. Freshwater wetlands provide a potential sink for atmospheric carbon, but, if not designed and managed properly, could become sources of greenhouse gases such as carbon dioxide and methane. According to published estimates of greenhouse gas fluxes from constructed and natural wetlands, fluxes from constructed wetlands are higher than those from natural wetlands, and the former have more carbon sequestration capacity than the latter. Wetland protection and restoration measures can improve the carbon sequestration potential of wetlands. However, it takes several decades for the carbon sequestration ability of restored wetlands to reach levels comparable to those of natural wetlands such as peatlands and forested wetlands.

Predicting how the carbon balance of wetlands will respond to anticipated climatic change requires a process-level understanding of carbon cycles through wetlands, mapping of the spatial distribution of relevant wetland characteristics, and the ability to predict how climate change will impact wetland hydrology and water depth. More research is needed to better understand the impacts of wetland water level fluctuations on carbon fluxes under variable climatic regimes. Further wetland research case studies should also aim to differentiate between methane production and consumption processes and evaluate their respective roles in carbon cycling and oxygen consumption both seasonally and during gradual system maturation.

The role of many wetland plants and microorganisms in carbon turnover and emitting methane is unclear. More research is needed to better understand the impacts of different plant species under variable nutrient regimes and loading rates. Carbon fluxes due to respiration via stems and leaves are not quantified for most wetland plants. The effect of temperature, oxygen penetration rate, and water column fluctuation on the methane oxidation rates and carbon turnover by microorganisms needs also to be clearly defined. Further work is required on methane fluxes for different wetland plant communities and associated microbial communities to find meaningful mechanistic relationships between processes such as methanogenesis and the biology of organisms responsible for the processes.

References

Alford DP, Delaune RD, Lindau CW (1997) Methane flux from Mississippi River deltaic plain wetlands. Biogeochemistry 37:227–236

Alongi DM, Trott LA, Pfitzner J (2007) Deposition, mineralization, and storage of carbon and nitrogen in sediments of the far northern and northern Great Barrier Reef shelf. Continent Shelf Res 27:2595–2622

Altor AE, Mitsch WJ (2006) Methane flux from created riparian marshes: relationship to intermittent versus continuous inundation and emergent macrophytes. Ecol Eng 28:224–234

Amthor JS, Dale VH, Edwards NT, Garten CT, Gunderson CA, Hanson PJ, Huston MA, King AW, Luxmoore RJ, McLaughlin SB, Marland G, Mulholland PJ, Norby RJ, O'Neill EG, O'Neill RV, Post WM, Shriner DS, Todd DE, Tschaplinski TJ, Turner RS, Tuskan GA, Wullschleger SD (1998) Terrestrial ecosystem responses to global change: a research strategy. ORNL Technical Memorandum 1998/27. Oak Ridge National Laboratory, Oak Ridge, TN

Anderson CJ, Mitsch WJ (2006) Sediment, carbon, and nutrient accumulation at two 10-year-old created riverine marshes. Wetlands 26:779–792

Armentano TB, Menges ES (1986) Patterns of change in the carbon balance of organic soil-wetlands of the temperate zone. J Ecol 74:755–774

Augustin J, Merbach W, Rogasik J (1998) Factors influencing nitrous oxide and methane emissions from minerotrophic fens in Northeast Germany. Biol Fertil Soils 28:1–4

Bano N, Moran MA, Hodson RE (1997) Bacterial utilization of dissolved humic substances from a freshwater swamp. Aquat Microb Ecol 12:233–238

Barber LB, Leenheer JA, Noyes TI, Stiles EA (2001) Nature and transformation of dissolved organic matter in treatment wetlands. Environ Sci Technol 35:4805–4816

Bartlett KB, Crill PM, Sebacher DI, Harris RC, Wilson JO, Melack JM (1988) Methane flux from the central Amazonian floodplain. J Geophys Res 93:1571–1582

Bedard-Haughn A, Jongbloed F, Akkerman J, Uijl A, Jong E, Yates T, Pennock D (2006) The effects of erosional and management history on soil organic carbon stores in ephemeral wetlands of hummocky agricultural landscapes. Geoderma 135:296–306

Bellisario LM, Bubier JL, Moore TR, Hanton JP (1999) Controls on CH_4 emissions from a northern peatland. Glob Biogeochem Cycles 13:81–91

Bormann BT, Spaltenstein H, McClellan MH, Ugolini FC, Cromackjr K, Nay SM (1995) Rapid soil development after windthrow disturbance in pristine forests. J Ecol 83:747–757

Boon PI, Lee K (1997) Methane oxidation in sediments of a floodplain wetland in south-eastern Australia. Lett Appl Microbiol 25:138–142

Brevik EC, Homburg JAA (2004) 5000 year record of carbon sequestration from a coastal lagoon and wetland complex, Southern California, USA. Catena 57:221–232

Bridgham SD, Richardson CJ (1992) Mechanisms controlling soil respiration (CO_2 and CH_4) in southern peatlands. Soil Biol Biochem 24:1089–1099

Bridgham SD, Megonial JP, Keller JK, Bliss NB, Trettin C (2006) The carbon balance of North American wetlands. Wetlands 26:889–916

Brix H, Sorrell BK, Lorenzen B (2001) Are *Phragmites*-dominated wetlands a net source or net sink of greenhouse gases. Aquat Bot 69:313–324

Bubier JL, Moore TR, Roulet NT (1993) Methane emissions from wetlands in the midboreal region of Northern Ontario, Canada. Ecol 74:2240–2254

Buesing N, Gessner MO (2006) Benthic bacterial and fungal productivity and carbon turnover in a freshwater marsh. Appl Environ Microbiol 72:596–605

Burgoon PS, Reddy KR, DeBusk TA (1995) Performance of subsurface flow wetlands with batchload and continuous-flow conditions. Wat Environ Res 67:855–862

Cao M, Gregson K, Marshall S (1998) Global methane emission from wetlands and its sensitivity to climate change. Atmosph Environ 32:3293–3299

Carroll P, Crill PM (1997) Carbon balance of a temperate poor fen. Glob Biogeochem Cycles 11:349–356

Carty A, Scholz M, Heal K, Gouriveau F, Mustafa A (2008) The universal design, operation and maintenance guidelines for farm constructed wetlands (FCW) in temperate climates. Biores Technol 99:6780–6792

Christensen TR (1993) Methane emission from Arctic tundra. Biochemistry 21:117–139

Christensen TR, Ekberg A, Ström L, Mastepanov M, Panikov N, Oquist M, Svenson BH, Nykanen H, Martikainen PJ, Oskarsson H (2003) Factors controlling large scale variations in methane emissions from wetlands. Geophys Res Lett 30:1–67

Clair TA, Warner BG, Robarts R, Murkin H, Lilley J, Mortsch L, Rubec C (1998) Canadian wetlands and climate change. In: Koshida G, Avis W (eds) Canada country study: climate impacts and adaptation. Environment, Ottawa, ON, Canada, Vol. VII: National sector volume, pp. 189–218

Collins ME, Kuehl RJ (2001) Organic matter accumulation in organic soils. In: Richardson JL, Vepraskas MJ (eds) Wetland soils. Genesis, hydrology, landscapes, and classification. Lewis/CRC, Boca Raton, FL, USA, pp. 137–162

Craft CB, Richardson CJ (1998) Recent and long-term organic soil accretion and nutrient accumulation in the everglades. Soil Sci Soc Am J 62:834–843

Crutzen PJ (1995) On the role of CH_4 in atmospheric chemistry: sources sinks and possible reductions in atmospheric sources. Ambio 24:52–55

D'Angelo EM, Reddy KR (1999) Regulators of heterotrophic microbial potentials in wetland soils. Soil Biol Biochem 31:815–830

Daulat WE, Clymo RS (1998) Effects of temperature and water table on the efflux of methane from peatland surface cores. Atmosph Environ 32:3207–3218

Davidson EA, Janssens IA (2006) Temperature sensitivity of soil carbon decomposition and feedbacks to climate change. Nature 440:165–173

Dick WA, Gregorich EG (2004) Developing and maintaining soil organic matter levels. In: Schjønning P, Elmholt S, Christensen BT (eds) Managing soil quality: challenges in modern agriculture. Centre for Agricultural Bioscience International, Cambridge, MA, USA, pp. 103–120

Euliss NH Jr, Gleason RA, Olness A, McDougal RL, Murkin HR, Robarts RD, Bourbonniere RA, Warner BG (2006) North American prairie wetlands are important nonforested land-based carbon storage sites. Sci Total Environ 361:179–188

Fey A, Benckiser G, Ottow JCG (1999) Emissions of nitrous oxide from a constructed wetland using a groundfilter and macrophytes in wastewater purification of a dairy farm. Biol Fertil Soils 29:354–359

Fischer H, Pusch M (1999) Use of the [14^C] leucine incorporation technique to measure bacterial production in river sediments and the epiphyton. Appl Environ Microbiol 65:4411–4418

Freeman C, Ostle N, Kang H (2001a) An enzymic "latch" on a global carbon store. Nature 409:149

Freeman C, Ostle NJ, Fenner N, Kang H (2004) A regulatory role for phenol oxidase during decomposition in peatlands. Soil Biol Biochem 36:1663–1667

Freeman C, Evans CD, Monteith DT, Reynolds B, Fenner N (2001b) Export of organic carbon from peat soils. Nature 412:785

Glatze S, Basiliko N, Moore T (2004) Carbon dioxide and methane production potentials of peats from natural, harvested and restored sites, Eastern Quebec, Canada. Wetlands 24:261–267

Gorham E (1991) Northern peatlands: role in the carbon cycle and probable responses to climatic warming. Ecol Appl 1:182–195

Gorham E, Underwood JK, Janssens JA, Freedman B, Maass W, Waller DH, Ogden JG (1998) The chemistry of streams in southwestern and central Nova Scotia, with particular reference to catchment vegetation and the influence of dissolved organic carbon primarily from wetlands. Wetlands 18:115–132

Grünfeld S, Brix H (1999) Methanogenesis and methane emissions: effects of water table, substrate type and presence of *Phragmites australis*. Aquat Bot 64:63–75

Haberl H, Erb K-H, Krausmann F, Adensam H, Schulz N (2003) Land-use change and socioeconomic metabolism in Austria. II: Land-use scenarios for 2020. Land Use Policy 20:21–39

Hanson RS, Hanson TE (1996) Methanotrophic bacteria. Microbiol Mol Biol Rev 60:439–471

Harriss RC, Sebacher DI, Day FP JR (1982) Methane flux in the Great Dismal Swamp. Nature 297:673–674

Hartel PG (2005) The soil habitat. In: Sylvia DM, Fuhrmann JJ, Hartel PG, Zuberer DA (eds) Principles and applications of soil microbiology, 2nd edn. Pearson/Prentice Hall. Upper Saddle River, NJ, USA, pp. 26–53

Hedges JI, Keil RG, Benner R (1997) What happens to terrestrial organic matter in the ocean? Organ Geochem 27:195–212

Hedmark Å, Scholz M (2008) Review of environmental effects and treatment of runoff from storage and handling of wood. Biores Technol 99:5997–6009

Hensel PF, Day JW Jr, Pont D (1999) Wetland vertical accretion and soil elevation change in the Rhône River delta, France: the importance of riverine flooding. J Coastal Res 15:668–681

Hobbie SE, Schimel JP, Trumbore SE, Randerson JR (2000) Controls over carbon storage and turnover in high-latitude soils. Glob Change Biol 6:196–210

Holden J (2005) Peatland hydrology and carbon release: why small-scale process matters. Philosoph Trans R Soc A 363:2891–2913

Hou AX, Chen GX, Wang ZP, Cleemput OV, Patrick WH (2000) Methane and nitrous oxide emissions from a rice field in relation to soil redox and microbiological processes. Soil Sci Soc Am J 64:2180–2186

Ibekwe AM, Grieve CM, Lyon SR (2003) Characterization of microbial communities and composition in constructed dairy wetland wastewater effluent. Appl Environ Microbiol 69:5060–5069

Inamori R, Gui P, Dass P, Matsumura M, Xu KQ, Kondo T, Ebie Y, Inamori Y (2007) Investigating CH$_4$ and N$_2$O emissions from eco-engineering wastewater treatment processes using constructed wetland microcosms. Proc Biochem 42:363–373

Joabsson A, Christensen TR (2001) Methane emissions from wetlands and their relationship with vascular plants: an Arctic example. Glob Change Biol 7:919–932

Joabsson A, Christensen TR, Wallein B (1999) Vascular plant controls on methane emissions from northern peatforming wetlands. Trends Ecol Evolut 14:385–388

Johansson AE, Gustavsson AM, Oquist MG, Svensson BH (2004) Methane emissions from a constructed wetland treating wastewater: seasonal and spatial distribution and dependence on edaphic factors. Wat Res 38:3960–3970

Johansson AE, Kasimir-Klemedtsson A, Klemedtsson L, Svensson BH (2003) Nitrous oxide exchanges with the atmosphere of a constructed wetland treating wastewater. Tell Series B Chem Phys Meteorol 55:737–750

Johnston CA (1991) Sediment and nutrient retention by freshwater wetlands: effects on surface water quality. Crit Rev Environ Control 21:491–565

Kadlec RH, Knight RL (1996) Treatment wetlands. CRC, Boca Raton, FL, USA

Kadlec RH, Knight RL, Vymazal J, Brix H, Cooper P, Haberl R (2000) Constructed wetlands for pollution control. Scientific and technical report no. 8, International Water Association, London

Kang H, Freeman C (2002) The influence of hydrochemistry on methane emissions from two contrasting Northern wetlands. Wat Air Soil Pollut 141:263–272

Kayranli B, Scholz M, Mustafa A, Hedmark Å (2010) Carbon storage and fluxes within freshwater wetlands: a critical review. Wetlands 30:111–124

Kelley CA, Martens CS, Ussler W (1995) Methane dynamics across a tidally flooded riverbank margin. Limnol Oceanogr 40:1112–1129

King GM (1996) In situ analyses of methane oxidation associated with the roots and rhizomes of a bur reed, *Sparganium eurycarpum*, in a Maine wetland. Appl Environ Microbiol 62:4548–4555

Kirschbaum MUF (1995) The temperature dependence of soil organic matter decomposition, and the effect of global warming on soil organic storage. Soil Biol Biochem 27:753–760

Krogh L, Noergaard A, Hermansen M, Greve MH, Balstroem T, Madsen HB (2003) Preliminary estimates of contemporary soil organic carbon stocks in Denmark using multiple datasets and four scaling-up methods. Agric Ecosyst Environ 96:19–28

Knight RL, Wallace SD (2008) Treatment wetlands. 2nd edn. CRC, Boca Raton, FL, USA

Knoblauch C, Zimmermann U, Blumenberg M, Michaelis W, Pfeiffer E (2008) Methane turnover and temperature response of methane-oxidizing bacteria in permafrost-affected soils of northeast Siberia. Soil Biol Biochem 40:3004–3013

Kragh T, Søndergaard M (2004) Production and bioavailability of autochthonous dissolved organic carbon: effects of mesozooplankton. Aquat Microb Ecol 36:61–72

Lafleur PM, Moore TR, Roulet NT, Frolking S (2005) Ecosystem respiration in a cool temperate bog depends on peat temperature but not water table. Ecosystems 8:619–629

Landry GM, Maranger R, Brisson J, Chazarenc F (2009) Greenhouse gas production and efficiency of planted and artificially aerated constructed wetlands. Environ Pollut 157:748–754

Le Mer J, Roger P (2001) Production, oxidation, emission and consumption of methane by soils: a review. Eur J Soil Biol 37:25–50

Li J, Wen Y, Zhou Q, Xingjie Z, Li X, Yang S, Lin T (2008) Influence of vegetation and substrate on the removal and transformation of dissolved organic matter in horizontal subsurface-flow constructed wetlands. Biores Technol 99:4990–4996

Liblik LK, Moore TR, Bubier JL, Robinson SD (1997) Methane emissions from wetlands in the zone of discontinuous permafrost: Fort Simpson, Northwest Territories, Canada. Glob Biogeochem Cycles 11:485–494

Liikanen A, Huttunen JT, Karjalainen SM, Heikkinen K, Vaisanen TS, Nykanen H, Martikainen PJ (2006) Temporal and seasonal changes in greenhouse gas emissions from a constructed wetland purifying peat mining runoff water. Ecol Eng 26:241–251

Limpens J, Berendse F, Blodau C, Canadell JG, Freeman C, Holden J, Roulet N, Rydin H, Schaepman-Strub G (2008) Peatlands and the carbon cycle: from local processes to global implications, a synthesis. Biogeosciences 5:1475–1491

Lloyd CR (2006) Annual carbon balance of a managed wetland meadow in the Somerset Levels, UK. Agric For Meteorol 138:168–179

Machate T, Noll BHH, Kettrup A (1997) Degradation of Phenanthrene and hydraulic characteristics in a constructed wetland. Wat Res 31:554–560

Maljanen M, Kohonen AR, Virkajarvi P, Martikainen PJ (2007) Fluxes and production of N_2O, CO_2 and CH_4 in boreal agricultural soil during winter as affected by snowcover. Tellus 59:853–859

Malmer N, Johansson T, Olsrud M, Christensen TR (2005) Vegetation, climate changes and net carbon sequestration in a North-Scandinavian sub-arctic mire over 30 years. Glob Change Biol 11:1895–1909

Mander Ü, Teiter S, Augustin J (2005) Emission of greenhouse gases from constructed wetlands for wastewater treatment and from riparian buffer zones. Wat Sci Technol 52:167–176

Mander Ü, Lõhmus K, Teiter S, Mauring T, Nurk K, Augustin J (2008) Gaseous fluxes in the nitrogen and carbon budgets of subsurface flow constructed wetlands. Sci Total Environ 404:343–353

McCarty GW, Ritchie JC (2002) Impact of soil movement on carbon sequestration in agricultural ecosystems. Environ Pollut 116:423–430

Mitra S, Wassmann R, Vlek PLG (2005) An appraisal of global wetland area and its organic carbon stock. Curr Sci 88:25–35

Mitsch WJ, Gosselink JG (2007) Wetlands, 4th edn. Wiley, New York

Moore TR, Dalva M (1993) The influence of temperature and water table position on carbon dioxide and methane emissions from laboratory columns of peatland soils. J Soil Sci 44: 651–664

Moore TR, Dalva M (1997) Methane and carbon dioxide exchange potentials of peat soils in aerobic and anaerobic laboratory incubations. Soil Biol Biochem 29:1157–1164

Moore TR, Roulet NT (1993) Methane flux: water table relations in northern wetlands. Geophys Res Lett 20:587–590

Moore TR, Roulet NT (1995) Methane emissions from Canadian peatlands. In: Lal R, Kimble J, Levine E, Stewart BA (eds) Soils and global change. Lewis, Boca Raton, FL, USA, Chap 12, pp. 153–164

Moore TR, Roulet NT, Waddington JM (1998) Uncertainty in predicting the effect of climatic change on the carbon cycling of Canadian peatlands. Clim Change 40:229–245

Moore TR, Bubier JL, Frolking SE, Lafleur PM, Roulet NT (2002) Plant biomass and production and CO_2 exchange in an ombrotrophic bog. J Ecol 90: 25–36

Mørkved PT, Dörsch P, Henriksen TM, Bakken LR (2006) N$_2$O emissions and product ratios of nitrification and denitrification as affected by freezing and thawing. Soil Biol Biochem 38:3411–3420

Ogden MH (2001) Atmospheric carbon reduction and carbon sequestration in small community wastewater treatment systems using constructed wetlands. In: Mancl K (ed) Proceedings of on-site wastewater treatment. 9th national symposium on individual and small community sewage systems, American Society of Agricultural Engineers, Fort Worth, TX, pp. 674–683

Öquist MG, Petrone K, Nilsson M, Klemedtsson L (2007) Nitrification controls N$_2$O production rates in frozen boreal forest soil. Soil Biol Biochem 39:1809–1811

Picek T, Cızkova H, Dusek J (2007) Greenhouse gas emissions from a constructed wetland – plants as important sources of carbon. Ecol Eng 31:98–106

Pind A, Freeman C, Lock MA (1994) Enzymatic degradation of phenolic materials in peatlands – measurement of phenol oxidase activity. Plant Soil 159:227–231

Pinney ML, Westerhoff PKM, Bakerm L (2000) Transformations in dissolved organic carbon through constructed wetlands. Wat Res 34:1897–1911

Price JS, Waddington MJ (2000) Advances in Canadian wetland hydrology and biogeochemistry. Hydrol Proc 14:1579–1589

Qualls RG, Haines BL (1992) Biodegradability of dissolved organic matter in forest throughfall, soil solution, and stream water. Soil Sci Soc Am J 56:578–586

Quanrud DM, Karpiscak MM, Lansey KE, Arnold RG (2004) Transformation of effluent organic matter during subsurface wetland treatment in the Sonoran Desert. Chemosphere 54:777–788

Raghoebarsing AA, Smolders AJP, Schmid MC, Rijpstra WIC, Wolters-Arts M, Derksen J, Jetten MSM, Schouten S, Damste JSS, Lamers LPM, Roelofs JGM, den Camp HJMO, Strous M (2005) Methanotrophic symbionts provide carbon for photosynthesis in peat bogs. Nature 436:1153–1156

Raghoebarsing AA, Pol A, van de Pas-Schoonen KT, Smolders AJP, Ettwig KF, Rijpstra WIC, Schouten S, Damste JSS, Op den Camp HJM, Jetten MSM, Strous M (2006) A microbial consortium couples anaerobic methane oxidation to denitrification. Nature 440: 918–921

Rasmussen PE, Albrecht SL, Smiley RW (1998) Soil C and N changes under tillage and cropping systems in semi-arid Pacific Northwest agriculture. Soil Tillage Res 47:197–205

Reddy KR, D'Angelo EM (1997) Biogeochemical indicators to evaluate pollutant removal efficiency in constructed wetlands. Wat Sci Technol 35:1–10

Reddy KR, Delaune RD (2008) Biogeochemistry of wetlands: science and applications. CRC/Taylor & Francis, Boca Raton, FL, USA

Roulet NT, Ash R, Quinton W, Moore T (1993) Methane flux from drained northern peatlands: effect of a persistent water table lowering on flux. Glob Biogeochem Cycle 7:749–769

Salm J-O, Kimmel K, Uri V, Mander Ü (2009) Global warming potential of drained and undrained peatlands in Estonia: a synthesis. Wetlands 29:1081–1092

Savage KE, Davidson EA (2001) Inter-annual variation of soil respiration in two New England forests. Glob Biogeochem Cycles 15:337–350

Scanlon D, Moore TR (2000) Carbon dioxide production from peatland soil profiles: the influence of temperature, oxic/anoxic conditions and substrate. Soil Sci 165:153–60

Schipper LA, Reddy KR (1994) Methane production and emission from four reclaimed and pristine wetlands of southeastern United States. Soil Sci Soc Am J 58:1270–1275

Schlesinger WH (1991) Biogeochemistry: an analysis of global change. Academic, San Diego, CA, USA

Schlesinger WH (1997) An analysis of global change. Academic, Harcourt Brace, San Diego, CA, USA

Scholz M (2006) Wetland systems to control urban runoff. Elsevier, Amsterdam, The Netherlands

Scholz M, Trepel M (2004a) Hydraulic characteristics of groundwater-fed open ditches in a peatland. Ecol Eng 23:29–45

Scholz M, Trepel M (2004b) Water quality characteristics of vegetated groundwater-fed ditches in a riparian peatland. Sci Total Environ 332:109–122

Scholz M, Harrington R, Carroll P, Mustafa A (2007) The integrated constructed wetlands (ICW) concept. Wetlands 27:337–354

Shepherd D, Burgess D, Jickells T, Andrew JS, Cave R, Turner RK, Aldridge J, Parker ER, Young E (2007) Modelling the effects and economics of managed realignment on the cycling and storage of nutrients, carbon and sediments in the Blackwater estuary, UK. Estuar Coastal Shelf Sci 73:355–367

Sherry S, Ramon A, Eric M, Richard E, Barry W, Peter D, Susan T (1998) Precambrian shield wetlands: hydrologic control of the sources and export of dissolved organic matter. Clim Change 40:167–188

Sleytr K, Tietz A, Langengraber G, Haberl R (2007) Investigation of bacterial removal during the filtration process in constructed wetlands. Sci Total Environ 380:173–180

Smith LK, Lewis WM, Chanton JP, Cronin G, Hamilton SK (2000) Methane emissions from the Orinoco River floodplain, Venezuela. Biogeochemistry 51:113–140

Stadmark J, Leonardson L (2005) Emissions of greenhouse gases from ponds constructed for nitrogen removal. Ecol Eng 25:542–551

Stern J, Wang Y, Gu B, Newman J (2007) Distribution and turnover of carbon in natural and constructed wetlands in the Florida Everglades. Appl Geochem 22:1936–1948

Stottmeister U, Wießner A, Kuschk P, Kappelmeyer U, Kastner M, Bederski O, Muller RA, Moormann R (2003) Effects of plants and microorganisms in constructed wetlands for wastewater treatment. Biotechnol Adv 22:93–117

Ström L, Christensen TR (2007) Below ground carbon turnover and greenhouse gas exchanges in a sub-arctic wetland. Soil Biol Biochem 39:1689–1698

Ström L, Ekberg A, Mastepanov M, Christensen TR (2003) The effect of vascular plants on carbon turnover and methane emissions from a tundra wetland. Glob Change Biol 9: 1185–1192

Tanner CC (2001) Plants as ecosystem engineers in subsurface-flow treatment wetlands. Wat Sci Technol 44:9–17

Tanner CC, Adams DD, Downes MT (1997) Methane emissions from constructed wetlands treating agricultural wastewaters. J Environ Qual 26:1056–1062

Teiter S, Mander Ü (2005) Emission of N_2O, N_2, CH_4, and CO_2 from constructed wetlands for wastewater treatment and from riparian buffer zones. Ecol Eng 25:528–541

Tietz A, Langergraber G, Watzinger A, Haberl R, Kirschner AKT (2008) Bacterial carbon utilization in vertical subsurface flow constructed wetlands. Wat Res 42:1622–1634

Tipping PW, Center TD (2002) Evaluating acephate for insecticide exclusion of *Oxyops vitiosa* (Coleoptera: Curculionidae) from *Melaleuca quinquenervia*. Florida Entomol 85:458–463

Trettin CC, Jurgensen MF (2003) Carbon cycling in wetland forest soils. In: Kimble J, Birdsie R, Lal R (eds) Carbon sequestration in US forests. Lewis, Boca Raton, FL, USA, pp. 311–328

Trenberth KE, Jones PD, Ambenje P, Bojariu R, Easterling D, Tank AK, Parker D, Rahimzadeh F, Renwick JA, Rusticucci M, Soden B, Zhai P (2007) Observations: surface and atmospheric climate change. In: Solomon S, Qin D, Manning M, Chen Z, Marquis M, Averyt KB, Tignor M, Miller HL (eds) Climate change 2007: the physical science basis. Contribution of working group I to the 4th Assessment Report of the Intergovernmental Panel on Climate Change. Cambridge University Press, Cambridge, UK, pp. 235–336

Tuittila ES, Komulainen VM, Vasander H, Nykanen H, Martikainen PJ, Laine K (2000) Methane dynamics of a restored cut-away peatland. Glob Change Biol 6:569–581

Turcq B, Cordeiro RC, Albuquerque ALS, Sifeddine A, Simoes Filho FFL, Souza AG, Abrao JJ, Oliveira FBL, Silva AO, Capitaneo JA (2002) Accumulation of organic carbon in five Brazilian lakes during the Holocene. Sediment Geol 148:319–342

Turetsky M, Wieder K, Halsey L, Vitt D (2002) Current disturbance and the diminishing peatland carbon sink. Geophys Res Lett 29:21-1–21-4

Updegraff K, Bridgham SD, Pastor J, Weishampel P, Harth C (2001) Response of CO_2 and CH_4 emissions in peatlands to warming and water-table manipulation. Ecol Appl 11:311–326

van Bochove E, Thériault G, Rochette P (2001) Thick ice layers in snow and frozen soil affecting gas emissions from agricultural soils during winter. J Geophys Res 106:23061–23071

Van der Peijl MJ, Verhoeven JTA (1999) A model of carbon, nitrogen and phosphorus dynamics and their interactions in river marginal wetlands. Ecol Modell 118:95–130

Vavrova P, Penttila T, Laiho R (2009) Decomposition of Scots pine fine woody debris in boreal conditions: Implications for estimating carbon pools and fluxes. For Ecol Manag 257: 401–412

Voelker BM, Kogut MB (2001) Interpretation of metal speciation data in coastal waters: the effects of humic substances on copper binding as a test case. Mar Chem 74:303–318

Vymazal J (2007) Removal of nutrients in various types of constructed wetlands. Sci Total Environ 380:48–65

Waddington JM, Rotenberg PA, Warren FJ (2001) Peat CO_2 production in a natural and cutover peatland: Implications for restoration. Biogeochemistry 54:115–130

Waddington JM, Roulet NT, Swanson RV (1996) Water table control of CH_4 emission enhancement by vascular plants in boreal peatlands. J Geophys Res 101:775–785

Walter B, Heimann M (2000) A process-based, climate-sensitive model to derive methane emissions from natural wetlands: application to five wetland sites, sensitivity to model parameters and climate. Glob Biogeochem Cycles 14:745–765

Weishampel P, Kolka R, King JY (2009) Carbon pools and productivity in a 1-km^2 heterogeneous forest and peatlandmosaic in Minnesota, USA. For Ecol Manag 257:747–754

Whalen SC (2005) Biogeochemistry of methane exchange between natural wetlands and the atmosphere. Environ Eng Sci 22:73–94

Whiting GJ, Chanton JP (2001) Greenhouse carbon balance of wetlands: methane emission versus carbon sequestration. Tellus 53:521–528

Williams CJ, Shingara EA, Yavitt JB (2000) Phenol oxidase activity in peatlands in New York State: response to summer drought and peat type. Wetlands 20:416–421

Wolf DC, Wagner GH (2005) Carbon transformations and soil organic matter formation. In: Sylvia DM, Fuhrman J, Hartel PG, Zuberer DA (eds) Principles and applications of soil microbiology, 2nd edn. Prentice Hall, Upper Saddle River, NJ, USA, pp. 285–332

Wuebbles DJ, Hayhoe K (2002) Atmospheric methane and global change. Earth-Sci Rev 57:177–210

Wynn TM, Liehr SK (2001) Development of a constructed subsurface-flow wetland simulation model. Ecol Eng 16:519–536

Xue Y, Kovacic DA, David MB, Gentry LE, Mulvaney RL, Lindau CW (1999) *In situ* measurements of denitrification in constructed wetlands. J Environ Qual 28:263–269

Yu Z, Apps MJ, Bhatti JS (2002) Implication of floristic and environmental variation for carbon cycle dynamics in boreal forest ecosystems of Central Canada. J Vegetat Sci 13:327–340

Yurova A, Lankreijer H (2007) Carbon storage in the organic layers of boreal forest soils under various moisture conditions: a model study for Northern Sweden sites. Ecol Modell 204: 475–484

Zhang JB, Song CC, Yang WY (2005) Cold season CH_4, CO_2 and N_2O from freshwater marshes in northeast China. Chemosphere 59:1703–1705

Zhu T, Sikora FJ (1995) Ammonium and nitrate removal in vegetated and unvegetated gravel bed microcosm wetlands. Wat Sci Technol 32:219–228

Zweifel UL (1999) Factors controlling accumulation of labile dissolved organic carbon in the Gulf of Riga. Estuar Coastal Shelf Sci 48:357–370

.

Chapter 4
Wetlands and Sustainable Drainage

Abstract While Chapter 2 focused on wetland systems for pollution control, this chapter concentrates on the combination of wetlands with sustainable drainage and flood control technology and planning. Particularly large retention basins, detention tanks, and alternative concepts for sustainable drainage are assessed. Section 4.1 introduces a rapid assessment methodology for the survey of water bodies including large wetland systems such as sustainable flood retention basins (SFRB). This novel and timely SFRB concept is funded and promoted by the European Union. Moreover, Section 4.2 provides a classification example for different Scottish SFRB, highlighting the dominance of current and former potable water supply reservoirs. Section 4.3 summarizes a new sustainable (urban) drainage system (SUDS); *i.e.*, a combined wetland and detention system. This SUDS technique could be combined with SFRB. Finally, Section 4.4 introduces the novel concept of integrating trees into SUDS design. The section shows that trees have the potential to reduce runoff volumes via retention, evapotranspiration, and interception, highlighting missed opportunities in traditional drainage design.

4.1 Rapid Assessment Methodology for the Survey of Water Bodies

4.1.1 Introduction

4.1.1.1 Background

The assessment and implementation of the general concept of sustainable flood risk management is an emerging challenge in environmental and water management. The concept is being further advanced by research at The University of Edinburgh (Scholz 2006a, 2007a, b; Scholz and Sadowski 2009) and The University of Salford, and it is recommended that these papers be consulted prior to implementing sustainable flood risk management methodologies.

The proposed guidance manual explains the underlying philosophy behind the SFRB categorization system and provides advice on determining the variables for water bodies including SFRB in the field. The outputs from this process are then analyzed utilizing the SFRB tools (Scholz and Sadowski 2009).

The European Union (EU) Flood Directive (2007/60/EC) is spearheading a move to sustainable flood retention in Europe. The directive requires that flood risk planning be completed on a catchment scale and aligns this with the preexisting Water Framework Directive (2000/60/EC) catchments and River Basin Districts. In particular, climate change is likely to increase the severity and frequency of flood events, thereby increasing the associated hazard. This may threaten some existing flood defenses that were designed and built prior to the identification of climate change as an issue; therefore, they may require modification to ensure their sustainability. The EU recognizes that member states may face significant challenges in implementing the Flood Directive and has responded with programs such as the strategic alliance for water management actions (SAWA), which aims to provide tools and guidance to aid the member states in implementation. In particular, SAWA is aimed at aiding North Sea area member states in the implementation of the Flood Directive.

The SAWA is a consortium of 22 partner institutions from Norway, Sweden, Germany, The Netherlands, and the United Kingdom, working together to produce a range of tools and guidance documents to assist in the implementation of the Flood Directive. With such a large and varied group, the Dutch philosophy of 'poldern', which means that all members work together to achieve a beneficial outcome, thus recognizing the contribution each member has to make, has been adopted. This philosophy promotes close collaboration and so supports the broad aims of the program. Flooding is a complex spatial planning issue and therefore requires a range of tools and approaches to solve the problem. Solutions are likely to require SUDS solutions applied on a small scale combined with SFRB used on a large scale.

The classification tool in this guidance manual provides a rapid screening method for water bodies and flood defense structures. It can accurately assess water bodies designed for flood and diffuse pollution control and can be applied as a rapid screening method to identify water bodies and impoundments that have the potential to be used as part of a sustainable flood risk management strategy.

4.1.1.2 Rationale for Rapid Survey Method

Existing survey methods for water bodies and catchment assessments are based on the ecology, chemistry, and hydrology of a catchment. The methods to determine these characteristics are time consuming and expensive (Watzin and McIntosh 1999). As these methodologies are predominantly ecological, their outputs tend to overemphasize the ecological status of a water body, and this can give rise to conflict in the case of flood defense impoundments with high flood return periods, because these basins become overgrown and often achieve a high biodiversity.

In some cases in Europe, this has resulted in expensive flood defense structures that cannot be used because flooding would damage the ecology (IUCN 2000). Such conflicts need to be resolved through impartial debate and discussion with an objective assessment of the structure, its design purpose, and current status. Many existing hydrological models do not consider the flood control potential of existing dams and impoundments to contribute to hydraulic management, though reservoir release from drinking water reservoirs to maintain river ecology is an established management practice (Montaldo *et al.* 2004).

A further aspect of sustainability in flood risk management should be to consider the existing flood defense infrastructure and impoundments that already exist within a catchment. Considering that many agencies will have to undertake assessments of their areas and objectively classify the flood defense potential of the existing infrastructure (SEPA 2007), a detailed, expensive investigation is not always going to be practical. The system outlined in this guidance manual has proven to be inexpensive, rapid, and reliable as an assessment tool for existing flood retention basin infrastructures such as most SFRB. The SFRB concept has evolved since 2006 and is based on the views of diverse international groups of engineers, landscape planners, and environmental scientists and has withstood detailed scientific scrutiny.

4.1.1.3 Manpower and Equipment Requirements

The philosophy behind this methodology is that it is rapid and inexpensive to apply within a catchment and therefore should not require expensive equipment or detailed measurements. The solution is a two-stage process of combining a desk study and a field visit. The desk study can provide an estimate of most variables using a standard personal computer with an Internet connection in less than 40 min. The site visit involves locating the water body, recording SFRB and catchment details using a digital camera, and assessing the SFRB variables visually. This typically requires 20 min per site. Eight or more sites can be assessed within a day of fieldwork, and the data gathered can be fed into the SFRB assessment tool (*e.g.*, Scholz and Sadowski 2009) to objectively categorize the surveyed structures.

A crucial feature of the proposed approach is that it should preferably be used by a multidisciplinary group of assessors. Ideally, the group should have different areas of expertise such as engineering, environmental science, hydrology, landscape management, and flood control planning. A team of two to three is ideal, as it promotes discussion and debate during the surveys. It is possible to apply the method with a single assessor; however, there is often a risk that the outcome of the assessment will be biased towards (his or her) particular discipline. The methodology has been demonstrated to be most effective when different disciplines are combined within the assessment team.

Therefore, the basic equipment required for the entire process is a personal computer with an Internet connection to carry out research, a digital camera to record the details of the catchment and the SFRB, and a 1:25,000 to 1:50,000 scale map of the survey area. A global positioning system (GPS) unit with <5 m error is a useful additional tool to allow geographical locations to be recorded, which can subsequently be used in digital modeling. A range of receivers are also available for the European Magellan GPS system, and typically these can achieve accuracies of between 1 and 2 m with postprocessing modules (GPS 2009). The higher the quality of information used in the process, the more reliable the outcomes will be.

4.1.1.4 Survey Template

The SFRB survey method is based on completing a site survey template, which contains a total of 40 variables. Details of these are provided in the next section with practical guidance on how to determine each of them. A vital aspect of the classification system is to assign an estimated confidence level to each of the variables as they are determined. The confidence value (%) is an estimate by the assessors of how accurately each variable has been determined and the confidence that they have in the determination.

The confidence value has been banded into high, medium, and low confidence levels. A high confidence value is typically one that has been measured or can be estimated with a very high degree of confidence based on knowledge and experience. The confidence value then assigned is between 61 and 100%. In cases where the confidence value is between 31 and 60%, additional investigations should be conducted to improve the confidence value. In cases where the confidence value assigned to a variable is ≤30%, the variable should be treated as missing. It has been found that assessors who use the system tend to assign confidence values in 5% increments. This is undesirable if sufficient expertise to undertake the assessment is available as it changes a 100-point to a 20-point scale.

In addition to the section on site characterization variables comprising details of the land types within a catchment, and details of the SFRB and its hydraulics, there is another section on bias and purpose that should be completed after the site visit. This section is potentially more subjective than the 40 basin variables and considers what the structure has evolved into and its current range of uses, and therefore takes into consideration the sustainability of the structure while recognizing the SFRB design purpose.

The survey template is only a guide for case studies in temperate and oceanic climates. It requires modification to be applied effectively in other climatic zones and to accommodate various scales of infrastructure. In particular, descriptions for variables such as annual rainfall and seasonal impact should be adjusted to the application area. Moreover, it is recommended that national weather and mapping data should be used where available to decide on landscape and climatic variables and the appropriate ranges for these parameters. As the assessment tool depends on the selection of the relative positions rather than on the numerical boundaries of the bins, it provides a relatively objective and consistent output, which can be used to facilitate stakeholder discussions and identify infrastructure, which has the potential to be used in sustainable flood risk management planning.

4.1.1.5 Sustainable Flood Retention Basin Typology

The suggested typology for SFRB is based on the views of a multidisciplinary and multinational team of scientists, engineers, and landscape planners (Scholz 2007a, b) and has been refined over time (Scholz and Sadowski 2009). This collaborative approach has resulted in six types of SFRB being identified as the minimum practical number to accommodate the variety of roles and modes of operation of this diverse group of structures and semi-natural water bodies. Most SFRB are used for the collection of river flow and runoff, which is slowed down and later released downstream resulting in discharge waves being flattened and discharge periods extended, mitigating potential flooding (Scholz 2007a, b).

Many retention basins perform additional tasks such as infiltration for ground water recharge, drinking water supply, diffuse pollution mitigation, enhancement of recreational benefits such as water skiing, bird watching, and fishing, and green space provision. In fact, some SFRB have even become Sites of Special Scientific Interest after years of neglect resulting in high biodiversity. The current multifunctionality of SFRB is largely what makes these structures sustainable. It is the resulting diversity of stakeholders that can lead to conflicts over the status and function of an SFRB and hinder successful sustainable flood risk management planning implementation.

The survey method combines hard scientific and engineering data such as the dimensions of dams and structures, and catchment land types with softer more holistic landscape and environmental variables characterizing the potential for diffuse pollution mitigation. This provides an impartial assessment of SFRB characteristics, which recognizes both the structure's design purpose and its current uses. The SFRB categorization methodology is therefore relevant to a wide range of stakeholders including flood risk management planners, engineers, local authorities, and community groups. It provides an impartial quantifiable and consistent assessment and should be used to identify infrastructure with the potential to contribute to flood risk management planning.

4.1.2 How to Use This Guidance Manual

This manual contains all the required information to allow for a comprehensive water body (particularly potential SFRB) assessment in a survey case study area. It is intended that the method should be applied at a catchment scale where the catchment boundaries are defined and the SFRB characteristics within the catchment are then identified.

Once the locations of a potential SFRB in a study area have been identified, a desk study for each basin is undertaken using all available sources of information. Variables are then estimated or measured and recorded on a survey form. The next step is to perform a site visit to confirm the measured and estimated variables from the desk study, and the survey form is subsequently completed and finalized. A short description of each of the variables is supplied below with information on how to determine each of the parameters. The entire process can be completed within approx. 1 h for most sites.

Once the numeric information on an SFRB has been collected, the data should be entered into any of the published SFRB categorization tools (*e.g.*, Scholz and Sadowski 2009), and this will then impartially and reliably categorize the surveyed structures and water bodies.

Scholz (2007b) outlines an empirical tool for the categorization of SFRB. A more statistical tool was developed for estimating SFRB types. This tool has been outlined by Scholz and Sadowski (2009).

4.1.3 Assessment of Classification Variables

4.1.3.1 Overview

The following sections provide a detailed overview of each of the water body classification variables and how to assess them. The definitions and advice are provided to ensure consistency of application of the SFRB determination method. Nevertheless, the author accepts that there may be specific case study sites where the information given below might be insufficient to determine a variable with high confidence.

4.1.3.2 Engineered (%)

A body of water can either be formed naturally or it can be created by man. Man-made structures can be highly diverse and range from very large water supply and hydropower dams of earth or concrete to small-scale structures often built to supply water to industrial processes. The *engineered* variable determines whether the SFRB is natural or man-made and how pronounced these tendencies are.

Dam structures with full engineering control, such as a drinking water supply reservoir or a purpose-built hydraulic flood retention basin, are examples of potentially highly engineered structures. These types of basin would typically receive values of between 70 and 95%.

A structure that is natural is one that has not been extensively modified by humans. The basin is typically a natural landscape feature and there is little (*e.g.*, formal outlet or protected embankment) or no evidence of human interference. This type of basin will normally receive a value of <20%.

The section on bias and purpose below has a strong influence on this variable, particularly in cases where the original design purpose and current use are different. Numerous industrial impoundments still exist in many areas, long after the industry they supported closed. These basins have become overgrown and neglected, often resulting in high biodiversity, and in other cases, these basins are now designated nature reserves (Johnson *et al.* 2003). These industrial relics now fulfill an entirely different role, and this is recognized in the division of scores between the different bias and purpose categories.

4.1.3.3 Dam Height (m)

Dam height is defined as the height of the man-made structure that creates the impoundment. The height of the structure is taken from the highest to the lowest point, usually just below the bottom outlet.

In some countries, databases of dams are available that contain details of dam structures and reservoir capacities. Where this type of resource is available, it is recommended that it be utilized to gather accurate information for the survey sites.

For a rapid assessment, the dam height can be visually estimated, and a valuable tool is a photograph with the face of the dam and a team member on top. The team member can then be used to scale the structure. Accurate GPS can also be used to record spatial coordinates at the top and base of the SFRB, though these can have high levels of error associated with them as well. Typically, the height has to be estimated from the front face of the dam as this is clearly visible.

4.1.3.4 Dam Length (m)

The *dam length* is defined as the span of the structure creating the impoundment. The structure may span a valley or, in rare cases, a dam may surround virtually the entire water body. Note that a natural lake restricted only by the topography of a valley attracts a dam length value of 0 m.

If written documentation of dam details is available, then this variable can be determined from this information. *Dam length* can be paced or measured during the site visit, or, for structures >500 m long, they can be measured from maps with a scale of ≤1:50,000.

4.1.3.5 Outlet Arrangement (%)

The outlet variable describes how and where water leaves the SFRB. In the case of small natural water bodies, there is usually a single river leaving the water body. This arrangement can be considered as a single, independent, simple, and uncontrolled outlet. The river outlet will not have any form of dam, weir, or sluice to control water levels and would generally receive a value of between 0 and 8% (with zero applied where there are no control structures). An entry of 8% might be given if a simple control or measurement structure was present but did not significantly impact water levels.

In the case of natural flood retention wetlands, the river outlet is typically natural and would receive a value of zero or close to it. The outlets in many shallow wetland systems can be choked by reeds and other vegetation, and these may slow down water release and treat it at the same time. This is one reason why these natural systems are considered SFRB.

Many relatively small dams and impoundments have a minimum of two types of outlets. Typically, a dam will have an overflow (spillway) and pipes within the dam structure for various purposes; *e.g.*, bottom outlet to drain the reservoir, outlet to control the river base flow, and outlet to convey the water for subsequent treatment. A complex outlet structure is usually considered to be a combined system (typically with manual operation and a single fixed spillway) and attracts values between 15 and 75%, with higher values being awarded to highly engineered systems with potentially fully automated and remotely controlled structures.

Large and modern drinking water reservoirs or hydropower dam constructions typically have combined outlets with fixed spillways and one or more water extraction points. These larger dams are often fully automated and typically receive a value higher than 75%.

4.1.3.6 Aquatic Animal Passage (%)

Key to the movement of aquatic animals is the flow of water from the basin and the geomorphology of the water course or overflow. In the case of many dams, the only possible route of movement to upstream areas for fish and other aquatic organisms would be up the spillway and outlet pipes. Typically, these are long concrete or masonry channels with significant drops in height and very thin layers of water making aquatic animal passage almost impossible (Larinier 2000). In the absence of a fish pass or fish ladder (Figure 4.1), these should be considered significant barriers to aquatic animal passage, typically receiving values between 0 and 10%.

In the case of some smaller SFRB, steps may or may not have been taken to facilitate aquatic animal passage. Generally, those SFRB with an adequate flow of water, no significant drops in height, and an insignificant barrier such as a small dam would score between 10 and 40%. Most SFRB that have been designed with a small fish pass or bypass stream would score between 41 and 69%.

Semi-natural water bodies and dams with a modern fish ladder are considered to allow for adequate aquatic animal movement from the impoundments to the wider environment, and typically these would receive values of >70%. Fish ladder design and passage has been a controversial issue for some time, and information on the effectiveness of fish ladders is a valuable aid in determining the value of this variable (Larinier 2000).

Figure 4.1 Typical Scottish
fish ladder

4.1.3.7 Land Animal Passage (%)

The *land animal passage* variable is intended to provide information on how easily terrestrial animals such as deer, squirrels, and birds can navigate across a dam
or around a water body. This variable requires consideration of the structure of the
SFRB and the wider landscape context along with any natural or man-made barriers such as the dam itself. The basin location, dam height and length, fencing,
gates, bridges, paths, and thickness and type of fringing vegetation can be important factors in this assessment.

Sustainable flood retention basins located in the remote or upper reaches of
catchments, where there is generally sparse population and infrastructure, typically
represent areas where terrestrial animal passage is good and attract values between
70 and 100%. It is possible for very large dams, often crossing steep valleys, to
pose a significant barrier to animal movement due to the high dam structure and
the large size of the impoundment it creates. Spillways often create a break in the
dam wall, which can be difficult for animals to cross. Sustainable flood retention
basins located in urban areas and near roads may pose a significant barrier to terrestrial animal movements (Shepherd *et al.* 2009); typically, these circumstances
result in values between 0 and 20%.

Disused water supply reservoirs are often used for fishing and other recreational activities. Moreover, they may even be designated nature reserves. Such
dams often have bridges crossing the spillways and are managed to remain in

a natural state, and paths and fencing are put in place to limit human disturbance and increase site safety. Some large natural water bodies can be barriers to animal movement, and in these circumstances a value between 21 and 69% is typically awarded.

4.1.3.8 Flood Plain Elevation (m)

Flood plain elevation is defined as the maximum additional height that the water rises above the normal height of the basin to reach the flood plain (if present) and is determined during the site visit. It is usually possible to estimate the normal water level of an impoundment or lake by the distribution of debris and water marks around the edges. These can be hidden by high water levels, which can make the variable difficult to estimate.

It should, however, be noted that the grass level is often a good indicator of maximum flood plain elevation as this plant cannot tolerate long periods of submergence. A clear line where the grass ends is often a good indicator of maximum flood water level.

In the case of many dams, there is a spillway present. In all such cases the site has a 0-m flood plain elevation as the spillway ultimately sets the impoundment's capacity. In some parts of Scotland, such structures have been seen with estimated water depths of between 0.1 and 0.3 m, and this observation suggests that a flooding depth of <0.3 m is appropriate. Flood plain elevations for other types of SFRB are typically in a range of between 0.3 and 3 m based on experience with some water bodies in the west of Scotland. Some of these water bodies flood most winters and achieve additional depths of at least 3 m.

4.1.3.9 Basin Channel Connectivity (m)

The *basin channel connectivity* considers how an SFRB is connected to its water inlets and outlets, and whether it provides a direct path for water flow during flood events. It is an estimate of how directly connected the SFRB is to its water supply and main drainage route, that is, whether the basin is online or offline. For an online structure, the entire inflow water stream fills and flows through the basin easily, while for an offline basin the flood water bypasses the impoundment via an additional channel. The distance between the bypass channel and the main stream bed flowing through the SFRB is called the basin channel connectivity.

An online SFRB will have a water inlet and outlet that are virtually part of the river system; *i.e.*, the river effectively flows straight through the SFRB or water body. Such SFRB and natural water bodies receive a score of 0 m.

In the case of some purpose-built offline SFRB, they may only receive water when the river reaches significant flood volumes. These basins are typically built for long-return-period flood events and are located adjacent to the river. Such basins are considered to be offline, and the distance of this offset is recorded because offset basins typically have a lower negative ecological impact on a river than those that are built online (Colin *et al.* 2000; Collins and Walling 2007).

4.1.3.10 Wetness (%)

The variable *wetness* has been added to aid in distinguishing between permanently flooded SFRB, such as drinking water supply reservoirs and industrial impoundments, and purpose-built SFRB, which are largely dry and are only flooded occasionally in response to major storm events, when they fulfill their flood control function.

Many current and former drinking water supply reservoirs (operated by Scottish Water) effectively run at their maximum design capacity, discharging down their spillways. Such SFRB typically receive a value of >90% depending on how much of the basin they occupy. Deep natural water bodies typically receive similar values.

Some shallow SFRB are silting up due to a range of factors including natural landscape processes such as siltation aggravated by eutrophication (McLemore 1988). In some Scottish locations such as the wider central Scotland area, this process is accelerated by the removal of arable farming restrictions in the catchment once a drinking water supply reservoir is no longer used for its design purpose. These basins are typically shallow and boggy with extensive fringes of dense reeds and other macrophytes, and they may therefore have very little open water. Such sites typically receive a value of between 10 and 74%, depending on the proportion of open water present.

Purpose-built SFRB designed for long-term flood events may only be partly flooded once or twice per decade. Such sites are typically dry basins with high levels of vegetation, and some may be used predominantly for other purposes such as recreation or farming. These dry basins typically receive a value of between 0 and 9%.

4.1.3.11 Proportion of Flow Within the Channel (%)

This variable describes the proportion of river water that will flow directly through an SFRB and is linked to the variable *basin channel connectivity*. If an offline SFRB is present, the mean proportion of flow through the additional channel needs to be considered during the estimation. An offline SFRB receives a value of <100%, while an online SFRB always receives a value of 100%.

In the case of disused impounds, these are often found at full capacity with water leaving via the dam spillway; the majority of the flow will be within the spillway channel and typically a score of 100% is awarded. Natural water bodies and wetlands typically have a single main channel or river draining them, and these also receive a score of 100%.

4.1.3.12 Mean Flooding Depth (m)

The *mean flooding depth* is a combined parameter that is determined by adding the mean additional depth due to an average flood to the mean depth of the basin. Many small impoundments, natural water bodies, and SFRB are very shallow basins typically with an average depth of <1 m. Such SFRB have relatively low flooding depths and are often well vegetated with dense stands of macrophytes such as reeds (Bayley and Guimond 2008).

Large basins such as some drinking water reservoirs, purpose-built SFRB, and some large industrial impoundments are designed to supply water at quantities of $100,000\,m^3/d$. Typically, the impoundments' overall capacity is many times this volume, and such SFRB and reservoirs can have depths of >20 m, typically reflected in a very high dam. For such basins and for some natural water bodies, flooding depths can be >3 m due to steep sides and constrained outlets.

4.1.3.13 Typical Wetness Duration (d/a)

Wetness duration is an estimate of the mean number of days during which the basin is wet within a given year. This variable has been added to distinguish between permanently flooded types of SFRB, such as some drinking water reservoirs and industrial impoundments, and purpose-built SFRB, which may only be flooded very occasionally and are therefore largely dry systems.

Natural water bodies, drinking water supply reservoirs, and other forms of large-scale impoundments are often permanently wet, typically receiving a value of close to 365 d/a. It is recognized that dams are periodically drained for maintenance and inspection. However, this does not detract from the predominantly permanently wet nature of these structures.

Many SFRB designed for long flood return events may receive a value of as low as 1 d/a (or even less). Typically, such impoundments are designed to deal with long return period flooding events such as 20, 50, or even 100 years (Mohssen 2008).

4.1.3.14 Flood Duration (d/a)

Flood duration is different from *typical wetness duration*, as it only considers the mean number of days in a given year that the SFRB is actually flooded rather than being wet. Typically, this variable is estimated from information based on rainfall patterns and the variable *seasonal influence* to arrive at a probable number. If information is available on water levels for an SFRB, this can be used to accurately determine the number of days of flooding.

In areas of high annual rainfall (≥2 m/a) and with even a moderate seasonal influence, there can be relatively frequent flooding events, and sites may be flooded as much as 20 to 30 d/a (*e.g.*, west coast of Scotland). In areas of low rainfall

($\leq$0.4 m/a) with a low seasonal influence such as Mediterranean climates, flooding is likely to have a very low frequency of $\leq$2 d/a.

4.1.3.15 Basin Bed Gradient (%)

The *basin bed gradient* is the mean slope of the basin from the main inlet to the main outlet points. Ordinance survey maps should be used to determine the elevation of the basin at each end, and the maps may also be used to determine other variables such as the length of the basin.

4.1.3.16 Mean Basin Flood Velocity (cm/s)

The *mean basin flood velocity* is defined as the average speed of the water traveling through the entire basin from inlet to outlet during a flood. An 'educated guess' is usually used to estimate this value with the support of other variables such as the slope of the basin. Other means of investigation are too expensive and time consuming for a brief investigation.

A high value for this variable is associated with purpose-built SFRB located in upland areas where heavy flooding occurs and where a large basin gradient is apparent. The value for such basins can be as high as 150 cm/s. In comparison, a typical value for basins in lowland areas is <15 cm/s.

4.1.3.17 Wetted Perimeter (m)

The *wetted perimeter* is the length of land and solid material that the water in the basin comes into contact with. Components that are included in the total length are the entire perimeter of the basin, any islands that are within the basin, and any vegetation (*e.g.*, tree trunks and reed stems) that is protruding through the surface of the water. The use of ordinance survey maps can help to roughly determine the wetted perimeter of large, deep, and geometrically simple water bodies.

A brief experiment can be undertaken to estimate the perimeter of reeds, *e.g.*, three small square frames of 10×10 cm should be placed around a representative section of reeds to obtain a composite estimate (Sutherland 2006). The wetted perimeter for a very small basin such as a SUDS pond could be less than 100 m. In contrast, the wetted perimeter for a much larger basin including islands and vegetation could have a value of over 10,000 m.

4.1.3.18 Maximum Flood Water Volume (m³)

The *maximum flood water volume* is reached when a basin is flooded to its maximum capacity and can retain no more water without it spilling over into another basin or catchment. The two main variables that someone should focus on when calculating the maximum flood water volume are the *mean flooding depth* of the basin and the *flood water surface area.*

The numerical value for this variable depends predominantly on the size of the water body. If the surface area is small, then the volume will also be small in comparison to a water body that has a much larger surface area. For upland areas, the water depth of an SFRB is relatively more important than the surface area, while the opposite is true for lowland areas.

4.1.3.19 Flood Water Surface Area (m^2)

This is the mean area of the water surface when the basin has been flooded. This information cannot be found on a map as the water surface on the map is often based on the maximum or mean depth. Therefore, an estimate of the *flood water surface area* has to be drawn onto a map, and the surface area should subsequently be calculated from this drawing.

Depending on the surrounding landscape, some flood water surface areas can be much larger than the existing mean surface area. This is particularly the case for lowland areas. In contrast, for many upland locations within steep valleys, the flood surface area remains fairly similar to that of the actual surface area of the water.

4.1.3.20 Mean Annual Rainfall (mm)

The *mean annual rainfall* (Bronstert *et al.* 2007) is the long-term average of the depth of rain that falls within the catchment area within a given year. This information cannot be gathered from a site visit alone. However, the value can be obtained from a database (*e.g.*, meteorological office) or local weather station. For the UK, the Flood Estimation Handbook CD-ROM contains exact rainfall data (CEH 1999).

4.1.3.21 Drainage (cm/d)

The *drainage* variable represents how efficiently water moves through the unsaturated zone of the soil and away from the basin. It estimates the mean distance at which water can drain through the unsaturated zone of soil over the course of a typical day.

Drainage can be estimated in a variety of ways with different accuracies. If the soil series around the basin can be identified, the drainage should be characterized from its known drainage properties. Groundwater vulnerability maps are widely available for EU countries, and these can be used to give a general indication of the drainage in an area. Areas where groundwater is considered vulnerable to pollution typically have excellent drainage, and there are few retarding reactions in the overlying soil. Therefore, these areas can be considered to have good drainage. Equally, areas of low groundwater vulnerability are typically those with poor drainage properties or where a layer of clay or other impermeable material underlies the soil (Ramchunder *et al.* 2009).

There are exceptions to these general conditions such as cases of extensive and deep organic soils, which protect groundwater from potential pollution or thin soils underpinned by hard igneous rocks. It is important to assess the soil while on site to confirm the initial desk study assessment.

4.1.3.22 Impermeable Soil Proportion (%)

Permeability is the ease with which a soil allows water to pass through it and is largely determined by the soil type surrounding a basin. Highly permeable strata such as sand and gravel easily allow water to pass through with little retardation. Such highly permeable strata typically contain <2% clay. In areas dominated by other soil types, there will be variable and significantly higher levels of clay, with the most impermeable soils being those with a high proportion of clay. Heavy clay soils generally have very poor drainage characteristics (Bah *et al.* 2009).

Soil properties can be estimated in the field, and there are a range of guidelines to estimate soil types. An easy-to-use example is the Soil Texture Fact Sheet (Brown 2003). The clay content is used in conjunction with the amount of rock present in the soil to arrive at an estimate of the overall impermeable proportion of the soil surrounding an SFRB.

Alternatively, soil series maps (Gerasimo 1969) can be used to determine the soil type and the proportions of sand, silt, and clay within that soil. These findings need to be compared to the soil in the field to determine the proportion of rock present.

4.1.3.23 Seasonal Influence (%)

The variable *seasonal influence* is a sum parameter that is easy to estimate, if general knowledge of regional weather conditions and the landscape topography is available. Climatic conditions have a pronounced effect on whether a basin is wet or dry, and the frequency of flooding. For example, in the UK, winter is wetter than summer. Most lowland areas around the Mediterranean can be considered to have little seasonal variation, and therefore SFRB located in these areas are considered to be subject to low seasonal influence (<20%). In contrast, in SFRB located in mountainous areas such as the Alps or in regions that are subject to wet climates such as the west coasts of Scotland and Norway, there may be a highly pronounced seasonal influence (>70%). Consequently, SFRB identified in these areas are considered to be subject to high seasonal variations (Dawson *et al.* 2008).

4.1.3.24 Altitude (m)

This variable is a measure of the altitude (elevation) at which the site is located and is simply determined from a spot height on a map or by using GPS equipment. The site elevation for a reservoir with a dam is taken at the bottom of the side of the dam facing upstream.

4.1.3.25 Vegetation Cover (%)

This variable refers to the basin and not to the corresponding catchment. Dry SFRB can be completely vegetated and, if maintained, are typically grass covered and achieve a vegetation cover value of between 20 and 70%. Mature vegetation such as a full basin covered with trees is associated with a value close to 100%. In contrast, no vegetation or tarmac will result in values close to zero. It is worth noting whether the SFRB is maintained and vegetated with short grass, which will not impede water flow through the basin, or if the basin is fully covered with mature vegetation, which may reduce its capacity and slow down the movement of flood water.

In the case of wet basins, the area of the basin occupied by emergent and floating plants is estimated during the site visit. The SFRB are highly diverse and some may contain values as low as 1% in the case of steep sided maintained drinking water reservoirs to >90% for basins that are silting up naturally and that are fully covered by mature reed stands.

4.1.3.26 Algal Cover in Summer (%)

This variable provides an estimate of the degree of phytoplankton growth in a wet SFRB and is a surrogate for the degree of eutrophication in that SFRB. It is easiest to estimate accurately during a summer site visit. Dry SFRB used solely for flood control purposes may have no potential for phytoplankton growth.

Water bodies that are rich in nutrients often undergo one or more extensive algal blooms in summer. In countries with nutrient-poor waters such as most upland areas of Scotland and Norway, it is unproblematic to estimate the likely potential for pollution-related blooms. It can be more challenging for some lowland areas in central Europe where waters are typically higher in nutrient content and support more extensive algal and surface macrophyte communities (Anderson *et al.* 2002).

An alternative method is the use of a Secchi disk, which is a white plastic disk (15.3 cm in diameter) on a weighted line with an accurate depth scale. The disk is simply lowered into the water and the depth at which it is no longer seen is recorded. The more eutrophic the water bodies, the shallower the depth at which the Secchi disk disappears. However, this approach is of limited value for highly eutrophic waters.

4.1.3.27 Relative Total Pollution (%)

Pollution is typically defined as the introduction by humans of substances or energy that can have a deleterious effect on the environment. In the context of this survey method, the variable relative total pollution is a measure of how impacted an SFRB is by predominantly diffuse agricultural and urban pollution, and it is largely a function of the breakdown of land types within a catchment and the way that the land is used.

Diffuse agricultural pollution is a significant problem in many areas of Europe. The pollution arises as a consequence of the normal arable and livestock application of fertilizers and agrochemicals as management tools (Stevens and Quinton 2009). Some of these chemicals are transported by surface and groundwater movement and can be captured within SFRB, particularly those designed to trap sediment. The degree of runoff and impact depends on farming practices, chemical application rates, and tillage practices. Livestock farming can result in significant inputs of nitrates and phosphates from the animals, and corresponding microbiological contamination can be a problem during storm events (Edwards and Withers 2008).

Old mines and associated spoil heaps and processing areas can be significant sources of pollution of the water environment. There is a substantial history of mining for most minerals and metals throughout the world. It is therefore valuable to consider whether there may be any water contamination from this source, which may affect the pollution status of a SFRB (Shepherd et al. 2009). The presence of a mine near an SFRB could result in a pollution assessment from anything from 2% for an old mine with no visible impacts to 100% for a site contaminated by acid mine drainage.

Industrial processes can be a source of pollution within catchments. In Europe and North America such facilities are closely regulated and monitored so that they do not exceed strict consent conditions (IPCC 2007). These facilities would be included in the assessment, and a low value of between 3 and 5% pollution would be associated with such a site. Old, derelict industrial facilities can be associated with significant land contamination problems, and such sites should therefore be considered to have a high pollution potential (Loures and Panagopoulos 2007).

Many SFRB including larger SUDS impoundments such as wetlands and retention basins can be found in urban areas. These SFRB typically have a tacit diffuse pollution control and mitigation function as they are designed to receive the first foul flush of contaminants associated with storm runoff. Typically, these contaminants such as polyaromatic hydrocarbons, mineral oil, and grease from road traffic are trapped in the basin, and once the basin is dry, these can be broken down by photolysis. Contaminants such as metals tend to build up in the sediments of these basins, and these can pose a problem for waste disposal in highly contaminated areas (Scholz 2006a).

The overall level of pollution of a basin is assessed by taking all of the factors outlined above into account. The site visit is particularly valuable in this regard as it shows land use practices and often reveals the industries present within a catchment.

4.1.3.28 Mean Sediment Depth (cm)

The *mean sediment depth* is the average depth of the sediment within an SFRB structure whether wet or dry. The sediment depth can simply be determined in dry basins by digging a shallow hole and subsequently assessing the sediment profile.

In permanently flooded SFRB, the sediment is not accessible, and an estimate of the sediment depth is therefore derived. Freshwater sedimentation rates are highly variable and depend on a wide variety of factors. For example, research has revealed that sedimentation rates for oligotrophic waters are as low as 0.16 mm per year, though these can increase to a range of between 5.6 and 11 mm per year, when pollution by phosphate and nitrate occurs (Moss 1988).

In the absence of published information, an estimate of the sediment depth can be made using expert judgment based on the sedimentation rates outlined above (depending on eutrophic status) multiplied by the number of years that the SFRB has been present. It should be noted that water supply reservoirs are generally located in areas where there are low maintenance costs associated with siltation and are managed to maintain the vegetation cover within a catchment.

It is standard practice in limnological investigations to express sedimentation rates as grams of sediment per meter per year. If the bulk density of the sediment is known, it is a simple matter to convert this to the depth of sediment (Ramos-Scharron and MacDonald 2005). The research team is currently actively working on the development of a better method for sediment depth estimation based on simple field measurement techniques.

4.1.3.29 Organic Sediment Proportion (%)

The proportion of organic matter within sediment is determined by complex inter-actions in the water environment of a catchment. Organic material is provided by terrestrial plants and animals, combined with the production from aquatic algae, plants, and animals within a wet SFRB to establish the overall production of the water body. This organic input is then metabolized by bacteria and sediment-dwelling invertebrates, which utilize a large proportion of the organic carbon as an energy source. The final organic proportion of the sediment is an interaction of these metabolic processes and the deposition of gravel, sand, silt, and clay from within the catchment (Kuhn and Diekmann 2003).

In upland areas with high rainfall, there are usually relatively low proportions of organic matter present in the sediment. This type of upland catchment tends to host oligotrophic water bodies with limited primary and secondary production, and the majority of the organic matter within the basin is cycled through the biota (Mitchell 1991). These types of catchments tend therefore to have low proportions of organic matter present within the sediment and would typically receive organic sediment proportion values of between 1 and 3%. An exception to this general rule can be where dense conifer plantations have acidified the water in an upland catchment. In these cases, refractory conifer leaf litter is a feature of all visible sediments (Heal 2001), and such sites should be considered to have a very high organic sediment proportion of >7%.

In lowland areas and in areas where the waters are oligotrophic to mesotrophic, sediments are considered to have relatively high organic matter content due to the increased productivity of such waters, and these are typically assigned values of between 7 and 15%. Where waters are eutrophic and highly productive there is often significant deposition of organic matter at the bottom of a permanently wet SFRB (Chung et al. 2009). These sites are considered to have a relatively high organic matter content of between 16 and 30%.

Hypereutrophic water bodies are those that are permanently polluted by sources of nutrients such as sewage discharges or from diffuse or point source pollution. These water bodies have a high primary production and regularly suffer from algal blooms.

When dead algae sink to the bottom of a water body, they tend to cause anoxic conditions, reducing the breakdown rate of organic materials in the sediments. Concurrently pollutants such as nitrate, phosphate, and ammonia are released into the water body. In such situations it is possible to find organic matter at between 30 and 60% of the sediment (Zhang et al. 2009).

Values >60% of organic sediment should not routinely be assigned to SFRB unless there is clear and compelling evidence that this is the case. For example, an exception is where an SFRB is surrounded by peat bogs. Peat bogs are decayed and water-logged sphagnum moss and other plant materials that have partially decomposed. This type of soil contains virtually 100% organic material (Ukonmaanaho et al. 2006).

The ideal solution to determine the organic carbon content of sediment is by direct measurement of a homogenized sample, typically achieved using an organic carbon analyzer, which heats the sample and converts the organic matter present to carbon dioxide that is then measured (Schumacher 2002). This measurement could be used directly in the SFRB classification system.

Due to the complexities of sediment dynamics and particular geochemical circumstances, the above boundary conditions may not be fully applicable to the survey area of interest. It is recommended that the survey team consider whether the boundaries proposed above are suitable for their survey area. Information on areas with sedimentation problems can be found in river basin management plans prepared by the European environmental agencies (EA 2009).

4.1.3.30 Flotsam Cover (%)

Flotsam is defined as debris and waste that is floating on the surface of a water body. It can include items such as debris from tress, rubbish thrown into the water by humans, or even abandoned boats or drowned cars. The principal objective of this variable is to determine if there is flotsam present that might restrict the flow of water out of an SFRB.

Levels of flotsam vary widely from upland catchments, where there may be virtually no flotsam present, to urbanized SFRB that have been used as dumping grounds for cars, shopping carts, or other man-made debris. Ultimately, the value assigned for flotsam cover should reflect the proportion of cover of the outlet by flotsam.

This variable is an indirect measure of flow restriction. Therefore, basins with outlets that have little flotsam cover receive low values (typically <5%), whereas basins that are full of flotsam such as some invasive plant species or a lot of human rubbish would receive high values if the outlet structure is mostly covered with flotsam.

An important aspect to verify during a site visit is whether there are screens present on overflows to retain fish, and whether these are clear of debris or obviously clogged. In cases where a fine mesh-type screen is in place to retain fish, these can become clogged with leaves or artificial items such as plastic bags. Clogged screens can raise water levels within a basin, taking it above its design capacity.

4.1.3.31 Catchment Size (km^2)

The *catchment size* is the area of land from which water feeds into an SFRB. Information on the catchments that feed individual dams can be found on national registers of these structures and if available should be used as a high-quality data source. In the absence of a national register of dams and reservoirs, the European environmental agencies have defined river basin districts and sub-catchments within the larger units (House 2009). These are a potentially valuable resource in the assessment procedure.

If a geographical information system of the survey area is available, the system can be used to define the likely catchment area for the SFRB based on the topography of the base map. The simplest method is to use commercially available digital map packages. Typically, these have the ability to translate the map into a three-dimensional terrain model. The area of a catchment can then be marked on the map by using the hills and streams to define a likely catchment area.

However, this procedure can be somewhat more problematic for lowland areas where there is a predominantly flat topography. Paper maps can be used in the same way, and the area is then directly estimated from the map. Errors are usually relatively small.

4.1.3.32 Urban Catchment Proportion (%)

Water quality within any catchment is the result of a complex series of interactions between the water source, water properties, catchment geochemistry, and inputs from human activities. As this is a rapid screening method, a range of land types that are considered to have different polluting properties have been defined for the temperate survey areas.

In some parts of the world, these land use categories are not appropriate and different land use types may need to be included. The most important feature of such substitutions is that the land type has a known runoff and pollution potential and is widely applicable within the survey area.

The urbanized proportion of a catchment, for example, is simply the area occupied by man-made structures such as roads, farms, and towns. The urban catchment is likely to be an important source of diffuse pollutants to the water environment and is found globally.

In the case of SFRB, which have a very high urban catchment proportion (>90%), it may be appropriate to replace the natural land use categories below with different types of urban development. Such a division could include light or heavy industry, industrial estates, retail, and residential areas as appropriate. Further research on defining different sub-categories is currently in progress.

4.1.3.33 Arable Catchment Proportion (%)

Arable land is defined as areas where crops are grown either for commercial agricultural purposes or subsistence farming. The type of crop is likely to vary with climatic conditions and weather and can be used for any type of farming where rows of crops are interspersed with bare soil. In some parts of the world, a separate category for rice farming and aquaculture will be needed.

A high proportion of arable area generally contributes greatly to the diffuse pollution of water bodies, so the *relative total pollution* of the basin and the *organic sediment proportion* might be high.

4.1.3.34 Pasture Catchment Proportion (%)

Pasture is land where animals are taken to graze. In temperate climatic zones such as northern Europe, pasture land is typically managed and consists of relatively short and dense grass that looks like a monoculture. It is given a separate category as it can be a significant source of nutrient and microbiological contamination of the water environment.

4.1.3.35 Viniculture Catchment Proportion (%)

Viniculture is the practice of growing vines to produce grapes, which are later made into wine. Many vineyards are on steeply sloping land and have relatively bare soils. These conditions can result in significant runoff and soil erosion, if badly managed (Casali *et al.* 2009).

In many areas of the world, viniculture is a common practice; however, this variable is not relevant to cool temperate climates such as Scotland and Norway. The *viniculture catchment proportion* may therefore be omitted from surveys where this land category is not relevant.

4.1.3.36   Forest Catchment Proportion (%)

This variable is simply the proportion of the catchment that is covered by predominantly managed forest, can easily be estimated or measured using maps, and should be ground-truthed during a site visit. In heavily forested catchments, it may be desirable to distinguish between natural woodlands and forestry plantations, particularly where forestry plantations can be a significant source of diffuse pollution and acidification (Nisbet *et al.* 2002).

This variable may overlap with the *natural catchment proportion*. This is particularly the case for unmanaged and natural forests.

4.1.3.37   Natural Catchment Proportion (%)

This final category of land use is intended to cover the remaining proportion of the catchment and is considered to be the land where there is no or minimal human interference. It may therefore constitute potentially remote grassland, scrubland, moor, and similar types of land.

Generally, basins located in upper lands or in deep valleys have high *natural catchment proportions*. In contrast, basins located in urban areas always have very little or no natural features. The natural catchment proportion variable can be assigned a value during the site visit when comparing its characteristics with those of the corresponding rather urban and forest catchment proportions, all of which can be obtained from maps. A high fraction of natural catchment also correlates positively with a low *organic sediment proportion*.

4.1.3.38   Groundwater Infiltration (%)

Groundwater is considered to be the water that lies beneath the saturated zone of the soil and is composed principally of surface waters and rainfall that has percolated through the soil and into the underlying rocks and typically intersects with water bodies such as SFRB and natural lakes (UK Groundwater Forum 2009). This parameter indicates the proportion of the water within an SFRB that comes from groundwater, and it can be a significant source of water for some cases outlined below.

In the case of purpose-built dry SFRB, there is virtually no groundwater infiltration. Moreover, former industrial impoundments and drinking water supply reservoirs are typically lined with an impermeable layer of clay and are therefore isolated from the surrounding groundwater. Both types of SFRB receive a value of <5%.

Some wet SFRB and natural lakes may receive a fair proportion of their water from groundwater. Such basins are typically shallow and, due to the groundwater flow, contain a small lake or pond within a larger basin. This is generally most apparent during the drier period of the year where springs can become visible. Such basins typically receive a value of between 5 and 10%, with 10% being considered a typical value for a natural lake or pond.

Basins may be encountered in some regions where groundwater comprises between 10 and 40% of the flow of water from the SFRB, and these are considered to have a high dependence on groundwater. Very high values for groundwater infiltration to a basin would be between 41 and 50%.

In special cases some SFRB may receive more than 50% of their water supply from groundwater. These systems usually have no stream as an inflow source. At the time of writing, the author is only aware of one such example where a Scottish SFRB is being kept permanently wet by receiving groundwater. This basin is located in an area where there is significant hexavalent chromium contamination. Hexavalent chromium is a human carcinogen by inhalation, and therefore dust arising from a dry SFRB could pose a hazard to human health. Utilizing the groundwater to keep the basin wet is a novel approach that minimizes this risk. Approximately 80% of the corresponding SFRB water budget comes from groundwater, with the other 20% being supplied by road drainage and direct rainfall.

4.1.3.39 Mean Depth of Basin

The *mean depth of the basin* is simply the average depth of the impoundment. In the case of dry SFRB, the depth should be recorded as the possible mean depth to aid in computing the flood water capacity of the basin.

In the case of permanently flooded SFRB, the corresponding depth has to be estimated. Excellent sources of information on water depths are fishermen, who regularly fish the SFRB. Anglers are a passionate and knowledgeable group, and if a fishing club has a lease on an SFRB, its members generally have extensive understanding of its history as well as physical characteristics, which can be invaluable to the surveying team.

For very large SFRB, there may be accessible sources of survey information to provide the average depth of a water body. Where this information is available, it often contains information such as maximum depth.

In the vast majority of wet SFRB, the mean depth of the basin will need to be estimated. An estimate can be derived based on the likely basin bed gradient and the maximum height of the dam, which is often close to the maximum depth of the impoundment, and by assessing the surrounding landscape.

4.1.3.40 Length of Basin (m)

The *length of the basin* is the distance from the two points of the basin perimeter that are furthest away from each other at normal environmental conditions (*e.g.*, no flooding). In an ideal case, this is the distance between the inlet and the outlet.

4.1.3.41 Width of Basin (m)

The *width of the basin* is the distance across the basin in normal environmental conditions and is ideally at right angles to the basin length. However, a mean value may be determined for unusual geometrical shapes.

4.1.4 Bias and Purpose

4.1.4.1 Overview

This bias and purpose section is intended to establish the roles that the SFRB was first designed for as well as its current uses. It should be clear that not all SFRB are used for their original purpose; for example, in Scotland there are many former industrial impoundments and disused drinking water supply reservoirs. These structures are typically somewhat neglected, which then results in a high biodiversity, making them attractive to the local communities who now use them as nature reserves or for other recreational activities such as walking, bird watching, and fishing. It is these evolutions and changes in function with time that are captured by this section of the survey method.

4.1.4.2 Dominant Hydraulic Purpose

A SFRB with a dominant hydraulic purpose is typically either wet or dry. Wet SFRB are purpose-built water storage reservoirs for drinking water or for the regular supply of water to industrial processes. Dry types of SFRB with a predominantly hydraulic function are those built for long return period flood events, which may be part of an integrated system of flood defense reservoirs as is seen in Baden, Germany (Scholz 2006a).

4.1.4.3 Drinking Water Supply

In the UK, there is an extensive network of current and former drinking water supply reservoirs, and these are a significant landscape feature in some areas. This reflects Scotland's reliance on surface waters to supply drinking water. In many other parts of the world, groundwater is a significant source of drinking water, and therefore surface water reservoirs may be relatively rare. In Germany, for example, approx. 80% of drinking water is supplied by groundwater, while 70% of UK drinking water supplies come from surface waters (EU 2007).

4.1.4.4 Production Industry

Many industries, from iron and steel production to chemicals to foodstuff manufacturing, require high volumes of water to maintain industrial production. Historically, many plants built their own water supply reservoirs, and in some cases these impoundments still exist many years after the industry they served has gone. It can be difficult to accurately identify such sites without reference to historic maps.

In the UK, the Ordnance Survey has made all their old maps available digitally, and these are a valuable source of information and freely available to researchers (*via*) the UK National Academic Data Centre (EDINA 2009). Industrial impoundments can be found by comparing the series of historic maps for an area of interest. If similar high-quality map data are available for other countries, they should be used as a definitive source of information. Regardless of how the information is obtained, any impoundments built for industrial use would receive a high percentage for this category (Scholz and Sadowski 2009).

4.1.4.5 Sustainable Drainage

An SFRB may be a large SUDS, where the purpose is to uphold BMP in terms of infiltration, water quality improvement, and sustainable resource management (CIRIA 2004), which is the fundamental basis of current thinking on sustainability and seen as the BMP. New SFRB are designed to be sustainable (CIRIA 2004). However, this may not have been a consideration with historic SFRB, which were built before the modern practice of sustainability evolved. It is, however, important to recognize that these older structures may now contribute to sustainable drainage either through hosting a wetland system or by providing retention capacity or, if silting up, by retaining sediments, which can be regarded as sustainable drainage, and a low value of between 0 and 20% can be assigned for older structures.

4.1.4.6 Environmental Protection

Some water bodies including SFRB may have environmental benefits and are even protected, because they are part of nature reserves or have received the designation Site of Specific Scientific Interest. These SFRB are mainly for the protection of animals, vegetation, and ecology. Normally, they will have signs indicating that there are protected species in the area or have warning signs indicating so.

In such cases the score assigned to environmental protection can be as high as 50%, with the other 50% of the bias and purpose assigned to drinking water supply (Section 4.1.4.3) as was the original design purpose of the SFRB.

4.1.4.7 Recreational Benefits

The main purpose of some water bodies may be to benefit the general public by providing recreational enjoyment including sports, walking, fishing, and bird watching. These SFRB can be easily identified as they are often managed; *e.g.*, fishing areas are provided with huts, which provide fishing supplies and check permits of anglers.

4.1.4.8 Landscape Aesthetics

Most natural water bodies and some SFRB, often located in parks, contribute significantly to the aesthetic value of a landscape. These watercourses are looked after by a local authority to maintain an aesthetically pleasing view for walkers and other user groups.

4.1.5 Presentation of Findings Using Geostatistics

The above variables characterizing impoundments can be used to produce maps for flood risk management purposes. This section briefly discusses some example maps for the wider central Scotland area. Two geostatistical techniques have been used for the display of the variables. Ordinary kriging provides the best linear unbiased estimations with minimum error variance and is the most commonly used type of kriging. In comparison, disjunctive kriging is a non-linear generalization of kriging. This estimation technique allows for the conditional probability that the value of a spatially variable SFRB characterization parameter is greater than a cutoff level to be calculated. A detailed discussion on spatial statistics is, however, not within the scope of this book.

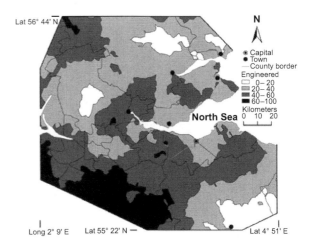

Figure 4.2 Map showing the application of ordinary kriging for the variable *engineered* (%)

Figure 4.3 Map showing the application of ordinary kriging for the variable *mean flooding depth* (meters)

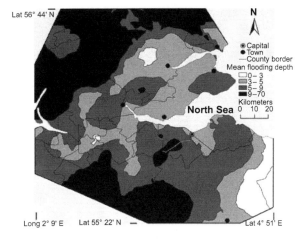

Figures 4.2 to 4.4 show map examples applying ordinary kriging for the variables *engineered, mean flooding depth*, and *maximum flood water volume*, respectively. High numerical values for the variable *engineered* generally indicate the likely necessity of high civil engineering investment to be made when planning for the construction of a new SFRB (Figure 4.2). The most engineered SFRB structures are likely to be found in the southwest of the study area, which coincides with the highest density of reservoirs and lakes used for water supply purposes. In contrast, low investment for flood infrastructure is required for the study area in the North. This variable is particularly useful when a decision has to be made on where an old flood infrastructure should be upgraded or a new SFRB should be constructed.

The spatial distribution for the variable *mean flooding depth* is shown in Figure 4.3. The mean flooding depth is relatively high in the less populated upland areas of the Northwest and South of the study area as well as within the Pentland

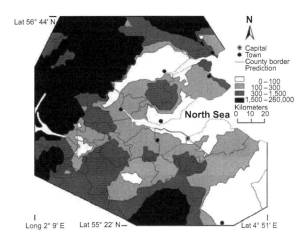

Figure 4.4 Map showing the application of disjunctive kriging for the variable *engineered* (%; exceeding 30%)

Figure 4.5 Map example showing the application of disjunctive kriging for the variable *engineered* (%)

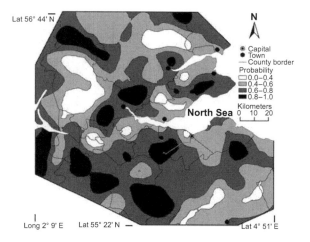

Hills area southwest of the capital, Edinburgh. Low values for mean flooding depth are rare and patchy.

Figure 4.4 shows the most likely values for the variable *managed maximum flood water volume*. This volume-based variable mirrors the depth-based variable, indicating that higher depths correlate with higher volumes, which is particularly the case for upland areas, far away from major urban settlements.

Map examples showing the application of disjunctive kriging for the variables *engineered, mean flooding depth*, and *maximum flood water volume* are summarized in Figures 4.5 to 4.7, respectively. Areas of low and high probabilities for the variable *engineered* are relatively small and patchy (Figure 4.5). This probability map can be used in conjunction with Figure 4.2 and all maps indicating flooding depth and flood water volume to determine the areas of greatest investment potential, if flooding is likely to be a problem.

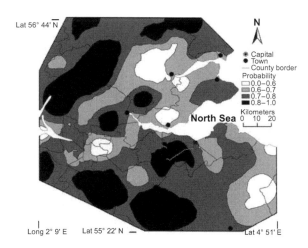

Figure 4.6 Map example showing the application of disjunctive kriging for the variable *mean flooding depth* (meters, exceeding 3 m)

Figure 4.7 Map example showing the application of ordinary kriging for the variable *maximum flood water volume* (m^3, exceeding $35 \times 10^{-4}\,m^3$)

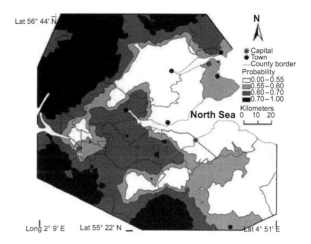

The map showing probabilities of exceeding 3 m flooding depth associated with the variables *mean flooding depth* should be used to estimate the likely return on flood infrastructure investment throughout the study area (Figure 4.6). The higher the probability, the more likely it is that an existing or planned SFRB will make a positive impact on flood control. The greatest potential for active flood control is in areas southwest of the capital such as the Pentland Hills. Figure 4.7 shows that the areas with the highest flood storage capacity are located in upland catchments far away from populated lowland areas.

4.2 Classification of Sustainable Flood Retention Basin Types

4.2.1 Introduction and Objectives

There are a wide range of traditional engineering solutions that can be applied to provide flood defenses in urban and rural areas. These traditional approaches utilize predominantly hard engineering solutions such as barriers and dykes to protect the public from the economic and social costs of flooding (Kendrick 1988). These traditional approaches are now supplemented by the availability of sustainable drainage systems, which generally operate by absorbing water and slowing down the rate of runoff from urban areas (Scholz 2006a). An emerging challenge for new and existing systems is climate change and the potential increase in rainfall and severe rainfall events that are expected to intensify in the future. Moreover, flooding often results in significant pollutant inputs to the water environment.

In light of this discussion, traditional flood retention basins have recently received increased attention by politicians, planners, and developers on the local and regional scale (Scholz 2007c). For example, the current design of German flood retention basins is based on outdated statistical rainfall events (ATV-DVWK 2001), which are now being called into question because of the reality of climate change.

Most natural and constructed retention basins keep runoff for subsequent release, thereby avoiding downstream flooding problems. Some basins such as wetlands do perform other tangible, albeit less 'visible', roles including diffuse pollution control and infiltration of treated runoff, promoting groundwater recharge. The diversity of retention basins is therefore high and further complicated by often multiple and competing functions that these structures fulfill.

A classification system is therefore needed to allow clear communication between stakeholders such as politicians, planners, engineers, environmental scientists, and public interest groups. The absence of a universal classification scheme for retention basins leads to confusion about the status of individual structures and their functions. This can lead to conflicts among stakeholders concerning the management of retention basins including wetlands. Therefore, Scholz and Sadowski (2009) proposed a conceptual classification model based on 141 SFRB located in the River Rhine Valley, Baden, Germany. Six SFRB types were defined based on the expert judgment of engineers, scientists, and environmentalists.

The European Union has acknowledged that member states may face significant challenges in complying with the flood directive and has therefore established programs such as the Strategic Alliance for Water Management Actions (2009) to develop guidance on adaptive measures such as SFRB to assist member states in developing flood risk management plans. This ongoing project will produce a database of adaptive structural and non-structural measures for the use of a wide range of stakeholders to aid them in the design of sustainable flood defense plans and to aid communication between the parties. It follows that a common classification system for water bodies, applicable across similar climatic conditions, would aid communication, planning, and understanding.

The aim of this section is to characterize sub-classes (*i.e.*, types) of SFRB in Scotland with the help of a revised rapid conceptual classification model, originally proposed by Scholz and Sadowski (2009). The key objectives are as follows:

- to aid stakeholder communication by avoiding misunderstandings with respect to planning and legal matters concerning the status of Scottish SFRB;
- to determine and characterize all relevant, and particularly the key independent, classification variables using a principal component analysis (PCA) and a sensitivity analysis applying the Wilcoxon test;
- to develop a universal conceptual classification methodology with the support of a large and detailed example case study data set; and
- to illustrate and discuss examples of the most dominant Scottish SFRB types that are also highly relevant for civil and environmental engineers and landscape planners.

4.2.2 *Methodology*

4.2.2.1 Identification of Sites and Definitions

One hundred and sixty-seven sites (Figure 4.8) were selected for classification using the 1:50,000 scale Ordnance Survey maps of central Scotland. In the context of this investigation, the sites of interest are those that may be able to play a role in either flood management or diffuse pollution control. Structures that may be able to play a role in flood control are considered to be those where the water level can be controlled either manually or automatically and are typically former or current engineered water supply reservoirs. Sites with the potential to contribute to diffuse pollution control are typically more natural and relatively small water bodies.

The most important classification variables for various types of SFRB in Scotland were identified and subsequently grouped (see below). Variables were deter-

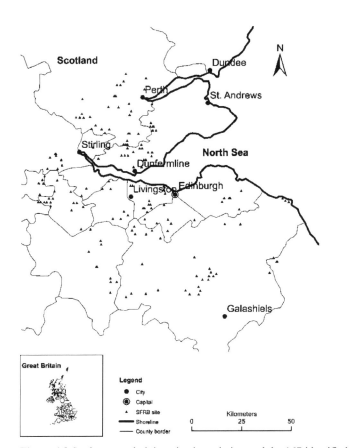

Figure 4.8 Study area, administrative boundaries, and the 167 identified sustainable flood retention basins in the wider central Scotland area

mined on the basis of literature reviews, various recent site visits in Germany, UK, Sweden, Ireland, and Denmark, and group discussions among British, German, Swedish, French, Irish, and American engineers, scientists, and landscape and urban planners.

The case study site investigation is a two-stage process comprising a desk study and site visit. During the desk study, the catchment boundaries for the water body, wet perimeter, area of water body, length of dam, elevation, basin gradient, and composition of catchment (urban, arable, forestry, and natural grassland proportion) are measured using digital maps. The site visit then ground-truths these findings, and the water body inflows and outflows are documented. Details regarding variables concerning the presence of a potential dam, its outlet control operation, basin catchment proportions, vegetation cover, and drainage are documented during a site visit and by collecting photographic evidence.

Table 4.1 Definitions for the sustainable flood retention basin (SFRB) types

Type	Name	Definition of SFRB types	Typical examples
1	Hydraulic flood retention basin (HFRB)	Managed traditional SFRB that is hydraulically optimized (or even automated) and captures sediment in a controlled manner	Drinking water reservoir (in operation); highly engineered and large flood retention basin (Figure 4.9)
2	Traditional flood retention basin (TFRB)	Aesthetically pleasing retention basin used for flood protection, potentially adhering to sustainable drainage practices and operated according to BMP	Former drinking water reservoir; traditional flood retention basin (Figure 4.10)
3	Sustainable flood retention wetland (SFRW)	Aesthetically pleasing retention and treatment wetland used for passive flood protection adhering to sustainable drainage and BMP	Sustainable drainage systems or BMP such as some retention basins, detention basins, large ponds, or wetlands (Figure 4.11)
4	Aesthetic flood treatment wetland (AFTW)	Treatment wetland for the retention and treatment of contaminated runoff that is aesthetically pleasing and integrated into the landscape and has some minor social and recreational benefits	Some modern constructed treatment wetlands; ICW (Figure 4.12)
5	Integrated flood retention wetland (IFRW)	Flood retention wetland for passive treatment of runoff, flood retention, and enhancement of recreational benefits	Some artificial water bodies within parks or near motorways that have a clear multipurpose function such as water sport and fishing (Figure 4.13)
6	Natural flood retention wetland (NFRW)	Passive natural flood retention wetland that may have become a Site of Specific Scientific Interest, potentially requiring protection from adverse human impacts	Natural or semi-natural lakes and large ponds, potentially with restricted access (Figure 4.14)

Figure 4.9 Hydraulic flood retention basin (sustainable flood retention basin type 1)

The user should be able to estimate most variables during a desk study, which should take approx. 20 min, and during a site visit of typically 40 min. A certainty percentage point (*i.e.*, low = 1 to 40%; medium = 40 to 60%; high = 60 to 100%) was attributed to each variable during the desk and field studies to reflect the likelihood of selecting a correct value. Certainty estimations depend very much on the expertise and bias of the user.

The authors' own revised definitions and characteristics for six sub-classes of SFRB as a function of their predominant purpose based on expert judgment, feedback from collaborators including landscape planners, data collected during desk studies, and field visits are listed in Table 4.1. Furthermore, the characteristics of

Figure 4.10 Traditional flood retention basin (sustainable flood retention basin type 2)

Figure 4.11 Sustainable flood retention wetland (sustainable flood retention basin type 3)

each SFRB type are also based on the interpretation of findings obtained from the statistical evaluation (see below). The six sub-classes are hydraulic flood retention basin (type 1), traditional flood retention basin (type 2), sustainable flood retention wetland (type 3), aesthetic flood retention wetland (type 4), integrated flood retention wetland (type 5), and natural flood retention wetland (type 6). The revised definitions of SFRB subclasses are independent of all statistical analyses and were formulated based on expert judgment based on empirical observations.

Figure 4.12 Aesthetic flood treatment wetland (sustainable flood retention basin type 4)

Figure 4.13 Integrated flood retention wetland (sustainable flood retention basin type 5)

Figure 4.14 Natural flood retention wetland (sustainable flood retention basin type 6)

4.2.2.2 Identification of Classification Variables

Data analyses were performed in Minitab (2003) unless stated otherwise. Most variables characterizing water bodies in Scotland were adopted from those initially proposed by Scholz and Sadowski (2009): 1. *engineered* (%); 2. *dam height* (m); 3. *dam length* (m); 4. *outlet arrangement and operation* (%); 5. *aquatic animal passage* (%); 6. *land animal passage* (%); 7. *floodplain elevation* (m); 8. *basin and channel connectivity* (m); 9. *wetness* (%); 10. *proportion of flow within channel* (%); 11. *mean flooding depth* (m); 12. *typical wetness duration* ($d\,a^{-1}$);

13. *estimated flood duration* (a^{-1}); 14. *basin bed gradient* (%); 15. *mean basin flood Velocity* (cm s^{-1}); 16. *wetted perimeter* (m); 17. *maximum flood water volume* (m^3); 18. *flood water surface area* (m^2); 19. *mean annual rainfall* (mm); 20. *drainage* (cm s^{-1}); 21. *impermeable soil proportion* (%); 22. *seasonal influence* (%); 23. *site elevation* (m); 24. *vegetation cover* (%); 25. *algal cover in summer* (%); 26. *relative total pollution* (%); 27. *mean sediment depth* (cm); 28. *organic sediment proportion* (%); 29. *flotsam cover* (%); 30. *catchment size* (km^2); 31. *urban catchment proportion* (%); 32. *arable catchment proportion* (%); 33. *pasture catchment proportion* (%); 34. *viniculture catchment proportion* (%); 35. *forest catchment proportion* (%); 36. *natural catchment proportion* (%); 37. *groundwater infiltration* (%); 38. *mean depth of the basin* (m); 39. *length of basin* (m); 40. *width of basin* (m).

Variables such as *engineered, floodplain elevation, basin and channel connectivity, mean flooding depth, flood duration,* and *relative total pollution* were refined and clarified to fit within the Scottish context. It has been appreciated that there are differences in the build environment and landscape. For example, the variable *mean flooding depth* recognizes high slope values for the Scottish landscape and deep flooding depths of some rather natural lakes.

The methodology has been updated by including the new variables *mean depth of basin, length of basin,* and *width of basin* in the classification template. The previous variable *aquatic and land animal passage* was divided into the following separate variables: *aquatic animal passage* and *land animal passage.* This accounts for fundamentally different obstacles concerning the freedom of unrestricted movement for animals.

Similarly, the old variable *forest and natural catchment proportion* was split into *forest catchment proportion* and *natural catchment proportion.* The former variable *viniculture catchment proportion* was not suitable for Scotland, so it was removed from the classification template.

The variable *wetness* was further refined to make a strong distinction between permanently wet systems such as reservoirs and lakes (Scottish data set) and SFRB, which may be dry and become wet only occasionally (German data set). *Typical wetness duration* became more important because it distinguishes between permanently flooded features such as reservoirs and lakes and SFRB designed for occasional flood control. The variable *flood frequency* is very difficult to determine with a high degree of certainty. Moreover, this variable is obsolete if no flood frequency data are available.

The variable *wetted perimeter,* also previously used by Scholz and Sadowski (2009), is highly important for the Scottish data set, which comprises predominantly wet basins in contrast to the dry basins dominating the German data set. A high *wetted perimeter* value is likely to indicate a higher diffuse pollution control potential. *Vegetation cover* has been further specified, considering that the vegetation within a predominantly dry basin is completely different to the aquatic vegetation within a wet basin.

Furthermore, industrial production and drinking water reservoirs were identified as new purposes for Scottish basins. These are in addition to the purposes of flood retention, sustainable drainage, environmental protection, recreation, and landscape enhancement identified for the German data set.

4.2.2.3 Rationale for the Elimination of Less Relevant Variables

The application of the PCA with the help of Matlab version 7.1 (Pratap 2002) helped to get a better overview of the underlying data structure. On the basis of the loading plot it is possible, where several variables are grouped closely together, to extract one single variable that may then replace the entire group. Besides the obvious time-saving advantages to this, the main point of the PCA is to remove redundant variables, hence reducing the risk of multicollinearity.

The cluster analysis and classification was performed twice, first using 39 variables and then using only 18 variables. Groups of variables formed by containing similar principal components were highlighted by circles around their corresponding labels. The final classification system was intended to be based on variables that are accurate and easy to obtain and that are associated with a high confidence value assigned to them during their determination.

Dominant variables were retained and used for a subsequent cluster analysis. The remaining variables within each group were discarded. This procedure has been followed because the use of too many variables may overcomplicate the decision-making tool, making the end product rather user-unfriendly. Furthermore, variables with similar principal components were effectively measuring the same fundamental variable. By keeping one variable representing a specific group, the other variables within this group naturally become redundant for the decision support tool.

Another technique to assess the suitability of a characterization variable is to evaluate its repeatability. Therefore, three groups assessed the same set of variables for 17 randomly selected case study sites independently of each other. The Wilcoxon signed rank test, also known as the Wilcoxon matched pairs test, is a non-parametric test used to assess the median difference in paired data. The test avoids the distributional assumption, because it is based on the rank order of the differences rather than the actual value of the differences. Non-directional hypotheses that there would be a significant difference between paired data (the initial site visit and the revisits) were made. The statistical analysis was carried out in the statistical package for the social sciences software (SPSS 2009) based on a two-tailed hypotheses.

4.2.2.4 Assignment of Sustainable Flood Retention Basin Types with the Help of Cluster Analyses

The statistical software package Matlab 7.1 (Pratap 2002) was used to perform cluster analyses on the standardized example data set. The clustering technique used was an agglomerative method (otherwise known as a bottom-up approach).

The results are displayed on a dendrogram, which allows an unambiguous appreciation of the cluster properties of the data.

The cluster analysis technique known as Ward's linkage, which effectively forces the data into a predefined number of clusters, thereby eliminating outliers, was applied (Kaufman and Rousseauw 1990). In this case, the objective was to obtain as many clusters as there are SFRB sub-classes, of which there are six.

After the Ward cluster analysis had grouped the 167 data point sites (one point corresponds to all 39 variable (excepting *viniculture catchment proportion*) values per site) into seven groups (six SFRB groups and one non-SFRB group), the general statistics of each cluster were found. The objective was to determine which SFRB type corresponded best to which newfound cluster, and this was done on the basis of expert judgment, supported by the case study information obtained during the site investigation. The dominant basin purposes greatly influence the selection of the most likely SFRB type; *e.g.*, a modern drinking water reservoir is likely to be a hydraulic flood retention basin (Table 4.1).

4.2.3 Findings and Discussion

4.2.3.1 Reduction Exercise for Classification Variables

An attempt was made to reduce the total number of variables based on the results of the PCA and a sensitivity analysis. The loading plot allowed 7 definite independent variables and 11 groups of dependent variables to be identified. For Scotland, *wetted perimeter, maximum flood water volume, flood water surface area, engineered, catchment size, outlet arrangement and operation, dam height, land animal passage, impermeable soil proportion*, and *mean sediment depth* were the most important independent SFRB characterization variables that greatly contributed to the variability expressed by the first and second components. The remaining 21 variables were regarded as redundant.

Dependencies were found for the following groups of variables: *engineered, dam height, outlet arrangement and operation*, and *impermeable soil proportion; flood duration and urban catchment proportion; drainage, vegetation cover*, and *relative total pollution; mean flooding depth* and *mean depth of basin; site elevation* and *natural catchment proportion; mean basin flood velocity, mean annual rainfall*, and *seasonal influence; pasture catchment proportion, forest catchment proportion*, and *groundwater infiltration; algal cover in summer, flotsam cover*, and *arable catchment proportion; aquatic animal passage* and *floodplain elevation; wetness, proportion of flow within channel, typical wetness duration*, and *organic sediment proportion; wetted perimeter, maximum flood water volume*, and *length of basin*.

It follows that variables with the following identification numbers are dependable: 2, 31, 24, 37, 23, 19, 34, 29, 7, 9, and 38. The remaining variables are redundant and could be omitted in the future. However, this observation is case specific.

4.2.3.2 Cluster Analyses

The cluster analysis was performed on all 39 variables and was based on the reduced set of 18 variables. The analysis performed on the reduced set of variables (independent, and easy and reliable to determine) indicated 7 clusters containing the 6 SFRB types and a group comprising non-SFRB sites (predominantly unmanaged natural lakes). Figure 4.15 shows the dendrogram for six SFRB based on 39 variables.

Concerning Figure 4.15, the clusters from left to right correspond to type 5 (group A; 16 sites), type 2 (group B; 57 sites), type 1 (group C; 4 sites), type 6 (group D; 68 sites), type 4 (group E; 9 sites), and type 3 (group F; 5 sites). Moreover, 8 sites were identified as non-SFRB. With respect to the analysis based on the reduced set of variables, the clusters correspond to SFRB type 6 (group A; 61 sites), type 3 (group B; 11 sites), type 5 (group C; 13 sites), type 4 (group D; 12 sites), type 2 (group E; 52 sites), and type 1 (group F; 5 sites). In addition, 13 sites were identified as non-SFRB.

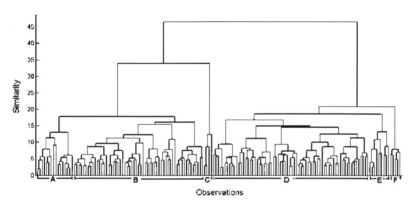

Figure 4.15 Dendrogram based on 39 variables for a data set of 167 retention basins (observations on x-axis) with Ward linkage and Euclidian distance used to identify the 6 sustainable flood retention basin types

4.2.3.3 Groupings Based on Cluster Analysis

Each cluster can be directly linked to an SFRB type, thus justifying their original choice, definition, and number. The distribution of cluster entries in the corresponding SFRB types was both explainable and expected. The reason is that virtually all artificial retention basins are initially built purely for flood protection or drinking water supply purposes. As a result, this purpose, and hence this SFRB type, still predominates even decades after construction or the last significant flood.

What has changed is that after years of having no major local floods, total dryness (or total wetness), or neglect, the purposes of many sites have changed, and the types have 'shifted' from the original purely hydraulic function to something more sustainable, aesthetic, or natural. Some sites have become so overgrown that they would no longer be able to handle the design flood and have instead become nature reserves, some even protected by law (usually type 6; Table 4.1). The conceptual model provides clear definitions for the past and current (*i.e.*, after aging) status of SFRB, thereby aiding communication among different stakeholders.

4.2.3.4 Application of Classification Methodology to Scotland

The number of classification variables was reduced with the help of a PCA. With respect to flood control, *dam length, basin bed gradient, flood water surface area, catchment size,* and *width of basin* were the most important independent SFRB characterization variables that greatly contributed to the variability expressed by the first and second component.

A cluster analyses was performed with a reduced set of variables that had been identified as independent and easy and reliable to determine. Seven clusters containing the six SFRB types and a group comprising non-SFRB sites (predominantly unmanaged natural lakes) were identified. The largest groups were natural flood retention wetlands (61 sites) and traditional flood retention basins (52 sites). The former includes passive natural flood retention wetlands characterized by a relatively high *wetted perimeter* and the latter comprises managed traditional reservoirs that are hydraulically optimized (partly automated). The relatively small groups represent SFRB that could also be classed as wetlands with strong flood and diffuse pollution control functions. Findings indicate that Scotland has a lower diversity of SFRB types than, for example, Baden (Germany), where six clear SFRB groups were identified.

This finding may have two principal reasons. The first reason may be bias in the selection of water bodies for investigation. However, this was not the case during this study, which was undertaken by a large and diverse team of experts over 3 years. The second reason is that there is a lower diversity of SFRB types in Scotland due to a simple or underdeveloped flood infrastructure that lacks retention structures overall.

The fieldwork program has identified a large number of water supply reservoirs that currently exceed requirements. In the vast majority of cases, these structures now fulfill multiple roles providing opportunities for recreation, nature conservation, and fishing, with many former drinking water or industrial water supplies being managed as fisheries.

A feature of these sites, based on the majority of current drinking water supply reservoirs surveyed, is that they are maintained at their maximum volumes and the spillways are continuously in operation. In this mode of operation, this extensive infrastructure contributes very little to water retention in the upper reaches of catchments.

It follows that a change in management practices of these structures could assist in sustainable flood risk management planning and result in more sustainable reservoirs. Effectively, this would require some water to be released from the reservoirs prior to expected heavy rainfalls. As the vast majority of former drinking water reservoirs have manual level control, this would require someone to visit the sites and open the valves to release the water, returning prior to the main rain-fall event to close the valves. This simple operation would create capacity to enhance water storage in the upper reaches of the catchments and retard the peak flows from the upper catchment, which has the potential to reduce the chances of flooding downstream. Combining this approach with conventional solutions such as sustainable drainage systems, barriers, and dykes will help to reduce the size, cost, and land take of other flood defenses. It is critical to the success of such an approach that appropriate compensation be provided to the owners of the structures to reflect the value of this service and the mild inconvenience it may cause.

As the most severe rainfall and storm events are usually predicted for the winter months, the reservoirs could be used for flood control purposes outside the fishing season. A major concern of the fishery owners will be the retention of the fish within the reservoirs during periods of water release, and this may require the fitting of screens onto the valve-controlled outlets of a reservoir. At the same time, water supply organizations such as Scottish Water will need to be reassured that the change of management practice will not impact negatively on the water quality within the basin and any management action would need to ensure that all the SFRB purposes and uses were maintained.

4.2.4 Conclusions

The Scottish data set contained only two main SFRB types. Traditional flood retention basins comprising predominantly former drinking water reservoirs are a visible component of the Scottish landscape. These structures could be used for low-cost flood control purposes, if their water level were actively controlled, which is currently not the case.

Natural flood retention wetlands also dominate the case study area and could make a significant contribution to diffuse pollution control, if managed appropriately. The most important independent and accurately determined SFRB variables that resemble wetland systems with a high diffuse pollution treatment function were *wetted perimeter*, *flood water surface area*, *engineered*, *catchment size*, *outlet arrangement and operation*, and *mean sediment depth*.

4.3 Combined Wetland and Detention Systems

4.3.1 Introduction

4.3.1.1 Background

Storm water runoff is usually collected in gully pots that can be viewed as simple physical, chemical, and biological reactors. They are particularly effective in retaining suspended solids (SS) (Bulc and Slak 2003). Conventionally, gully pot liquor is extracted on virtually random occasions from road drains and transported (often over long distances) for disposal at sewage treatment works (Butler *et al.* 1995; Memon and Butler 2002).

Storm water management strategies generally involve controlling non-point source pollution by implementing BMP (Olding *et al.* 2004; Wu *et al.* 2006). Runoff pollution has been characterized, in magnitude and in concentration of pollutants, by intermittent and impulse-type discharges into receiving waters, causing shock-loading problems for the ecosystems of these water bodies (Wu and Ahlert 1978; Ellis 1991).

Storm water detention systems treat runoff, for example, from parking lots and local roads and are frequently more environmentally sustainable in comparison to most traditional drainage technologies. This can reduce the costs of construction, transport, and treatment significantly. Moreover, other studies suggest that treated runoff can be used for irrigation purposes (Scholz 2006a).

Belowground storm water detention systems are defined as sub-surface structures designed to accumulate surface runoff and from where water is released, as may be required to increase the lag period of the rainfall and runoff hydrograph. The structure may contain aggregates with a high void ratio or plastic crates (also with a high void ratio; typically 95%) wrapped in geotextile and also act as a water recycler or infiltration device (Butler and Parkinson 1997).

Since 1980, belowground storm water detention systems have been installed that are specifically designed to reduce storm water flow. The surface water can be captured through infiltration or a distribution system comprising pipes or swales. The filtered storm water is detained below the ground within a detention tank (Butler and Parkinson 1997).

Concerning detention systems used simultaneously as soakaways, the runoff is often treated by filtration prior to infiltration or discharge to the sewer or watercourse via a discharge control valve. The application of these systems reduces runoff in case of minor storms as well as encourages groundwater recharge and pollution reduction. These novel detention systems can frequently be found in new developments (Scholz 2006a).

There is a need to modify common storm water detention systems based on belowground plastic crates to meet more stringent water quality guidelines (Butler and Parkinson 1997; Scholz 2006a). Further research should therefore also focus on the implementation of sustainable filters within the current structures of detention systems.

4.3.1.2 Microbial Contamination

Fecal pollution within storm water runoff can cause significant health risks as a result of the presence of various infectious microorganisms such as *E. coli* (Brion and Lingireddy 2003; Ellis 1991; He and He 2008; Kay *et al.* 2005;

Figure 4.16 Dog fouling

McCorquodale *et al.* 2004; Scholz 2006a). It is well understood that dog fouling (Figure 4.16) is the major source of fecal contamination in urban runoff. However, relevant risk studies on this subject have not been performed. The UK's dog population is reported to be between 6.5 and 7.4 million, producing nearly 1000 tons of feces per day. Additionally, daily fecal output per dog is estimated to range between 100 and 200 g (O'Keefe *et al.* 2005).

Bacterial indicator organisms have been frequently used to assess the presence of fecal contamination and, consequently, pathogens in drinking and bathing waters (NRC 2004). Total coliforms (TC) and *Enterococcus* are the most commonly used indicators (Ellis 1991; NRC 2004) due to their relative ease of application and low determination costs (Brion and Lingireddy 2003; Kay *et al.* 2005).

Modeling microbial water quality can be a useful approach for watershed managers, environmental regulators, and others involved in the evaluation and protection of ecological habitats and public health. Artificial neural networks (ANN) can be used to derive relationships between gathered data to predict microbial populations and other water quality parameters (Brion and Lingireddy 2003; Lee and Scholz 2006a).

The microbial population in storm water runoff is controlled by different variables including temperature, availability of SS, and nutrients. Studies show that enterococci preferentially attach to particles with diameters from 10 to 30 μm, while TC have a broader distribution (Jeng *et al.* 2005).

4.3.1.3 Modeling Approaches

An ANN can simply be described as an artificial computational copy of a brain (Iyengar and Kashyap 1989; Mohanty *et al.* 2002; Lee and Scholz 2006). The networks (operate) by attempting to mimic the way in which human brains function (Zurada 1992). Mathematically, an ANN is a non-linear function comprising parameters that can be trained by an optimization procedure so that the ANN output becomes similar to the measured output on a known data set (Scholz 2006a). This ability to replicate non-linear relationships makes ANN suitable for modeling environmental systems (Maier and Dandy 1998). Recently, ANN models have been used in many water resource applications such as water quality forecasting, the prediction of chemical dosages, and microbiological numbers in urban runoff and water treatment plants (Brion and Lingireddy 2003; He and He 2008; Maier and Dandy 2000; Neelakantan *et al.* 2001; Lee and Scholz 2006).

Frequently, ANN modeling approaches have been applied in the area of water quality modeling, where they proved to be particularly successful in predictions based upon complex, interrelated, and often non-linear relationships among multiple parameters (Sandhu and Finch 1996; Brion and Lingireddy 2003).

An ANN addresses complexity through a vast number of highly interconnected processing elements (called nodes in this chapter), working in concert to solve specific problems including forecasting and pattern recognition. Each node is connected to other neighboring nodes with different coefficients or weights, which represent the relative influence of the varying node inputs to other nodes (Hamed *et al.* 2004).

4.3.1.4 Aim and Objectives

The aim of this study is to show how different storm water detention systems cope with a 'worst-case pollution scenario'. The research objectives are to:

- assess the general inflow and outflow water quality;
- evaluate the water treatment efficiencies of different experimental storm water detention systems receiving concentrated runoff contaminated by dog feces;
- develop multiple regression models for each system;
- undertake ANOVA to compare inflows and outflows and the performances of all systems; and
- predict TC and intestinal enterococci colony unit formation by developing an ANN for each system and each variable.

4.3.2 Methodology

4.3.2.1 Experimental System Set-up

Five biologically (in terms of age) mature detention systems (Figure 4.17; plastic crates wrapped in geotextile, sponsored by Atlantis Water Management and Al-

Figure 4.17 Experimental set-up showing five different experimental detention systems

derborough, Sladen Mill Industrial Complex, Littleborough, UK), were located outdoors at The King's Buildings campus (The University of Edinburgh, Scotland, UK) to assess the system's performances under realistic natural conditions during a period of more than 1 year (1 April 2005 to 13 September 2006). However, the rig was in operation in batch flow mode since 31 March 2004. A practical application of the crate-based detention system is shown in Figure 4.18.

Two plastic crates (total height 1.7 m, length 0.68 m, width 0.41 m) on top of each other comprised one detention system. The tank volume below each filter was 0.08 m^3. The detention system filter volumes for all five systems were 0.24 m^3. All systems were located above ground to allow for easy accessibility. Due to this

Figure 4.18 Practical application of plastic crate-based detention system

arrangement, higher seasonal mean and more variable overall temperatures than usual were noted. However, this arrangement helped to assess a worst-case scenario regarding potential microbial regrowth due to higher temperatures, particularly during the summer.

The bottom cell (almost 50% full at any given time) was used for water storage and passive treatment only. The top cell was used as a coarse filter. Different arrangements of aggregates and planting were used within the filtration zones of each detention system. Different packing order arrangements of aggregates and plant roots were used in the systems to test for the effects of gravel, sand, sand mixed with bark, block paving, and turf on the water treatment performance.

Systems 1 and 2 represented sand, and gravel-filled constructed wetlands planted with *Phragmites australis* (Cav.) Trin. ex Steud (Common Reed), and a detention basin, respectively. Systems 3, 4, and 5 were similar to slow sand and trickling filters.

Inflow water, polluted by road runoff, was collected after mixing by manual abstraction with a 2-l beaker from randomly selected gully pots on the campus and nearby roads. This is a standard approach as discussed by Scholz (2006) and Lee and Scholz (2006). Temperature and dissolved oxygen were measured onsite, and the corresponding water samples were subsequently transferred into the campus-based public health laboratory for further water quality analyses.

All detention systems were watered as slow as possible within 3 to 5 min approx. twice per week with 5-l gully pot liquor artificially contaminated by dog feces (180 g) and drained by gravity afterwards to encourage air penetration through the filtration system (Gervin and Brix 2001). The amount of dog feces chosen was based on data obtained during an unpublished internal survey of dog fouling around The King's Buildings campus. The quantity of gully pot liquor used per system was approx. 3.6 times the mean annual rainfall volume (data obtained from The University of Edinburgh Weathercam Station in 2006) to simulate a 'worst-case scenario' for a hypothetical catchment similar to the wider campus area. The hydraulic residence times were in the order of 1 h.

4.3.2.2 Data Set

The sampling of data was done simultaneously for all systems. However, the number of samples is sometimes different between inflow and outflow for the same variable, because outliers and human error are identified at a later stage during data analysis. Consequently, values identified as flawed have been removed with the help of the box plot method from the data set. It follows that correct data that directly correspond to all removed entries were also removed during further analysis and modeling to obtain an overall data set that only contains matching pairs. All tested variables were $\log_{10}$-transformed to achieve normality for subsequent statistical tests, if required.

4.3.2.3 Modeling

There are difficulties involved with applying ANN models for microbial water quality predictions, mostly as a result of complexities in environmental distribution and because of the mobility and fate of microbes. Microbial contaminants are known to be non-conservative and unevenly distributed, and their numbers and growth rates may change in the environment depending on the conditions they live in. The interrelationship and interactions between microbial colonies in storm water cause various modeling challenges that have been overcome for particular case studies by applications of ANN to multiparameter data sets (Brion and Lingireddy 2003; He and He 2008).

Each neuron in an ANN has a scalar bias b, the bias is similar to a weight except that it has a constant input of 1. The transfer function net input n in the ANN is also a scalar and is equal to the sum of the weighted input wp and the bias b. This sum is the argument of the transfer function f. A transfer function can be a step function or a sigmoid function that takes the argument n and produces the output a. Both w and b are adjustable scalar parameters of the neuron. The main concern in ANN is the adjustability of such parameters so that the network would be able to reveal the most desired and interesting behavioral patterns.

ANN vary in type. A basic neural network contains one input, one hidden, and one output layer. These layers are all connected without any feedback connections. The weighted sum of the inputs are transferred to the hidden nodes, where it is transformed using an output function (also called transfer or activation function). In return, the outputs of the hidden nodes perform as inputs to the output node where another transformation happens. Network outputs often have associated processing functions; these functions are used to transform user-provided target vectors for network use. Network outputs are then reverse-processed using the same function to produce output data with the same characteristics as the original user-provided targets. The processing function for the output of the hidden layer is the output function given in Equation 4.1:

$$x_i = \sigma_i\left(b_i + \sum_{j=1}^{n} w_{ij} u_j \right),$$

(4.1)

where x_i is the output from the hidden node, σ_i is the output function of the hidden node (usually the hyperbolic tangent $tanh$), b_i is the bias input to hidden node i, n is the number of input nodes, w_{ij} is the weight connecting input node j to hidden node i, and u_j is input node j (Sarle 2002).

The processing function for the output of a network can be expressed in Equation 4.2:

$$y_i = \sum_{j=1}^{m} c_{ij} x_j,$$

(4.2)

where y_i is the output from output node i, m is the number of hidden nodes, c_{ij} is the weight connecting the hidden node, and x_j is the weighted sum of inputs into hidden node j to output node i.

During network training, the connection weights and biases of the ANN are adapted through a continuous process of simulation. The primary training goal is to minimize an error function by searching for connection strengths and for biases that make the ANN produce outputs that are equal or close to the targets. Equation 4.3 expresses the mean square error (MSE) of the output values:

$$MSE = \sum_{t=1}^{N} \left(Y_t - \widehat{Y}_t \right)^2 / N, \tag{4.3}$$

where MSE is the mean square error, N is the number of data points, Y_t is the observed output value, and $\widehat{Y}_t$ is the output of a feedforward neural network.

The minimization procedure consists in the optimization of a non-linear objective function. A number of optimization routines can be applied. For practical purposes, the Levenberg–Marquardt routine is often used as it finds better optima for various problems than the other optimization methods (Sarle 2002).

4.3.2.4 Development of the Artificial Neural Network Model

In this study, one of the most commonly used types of ANN was used: the feedforward network, where the information is transmitted in a forward direction only. According to Tomenko *et al.* (2007), feedforward ANN models were found to be one of the most efficient and robust tools in predicting constructed treatment wetland performance compared to traditional models.

A multilayer, feedforward ANN usually contains one input, one output, and one hidden layer. Different numbers of hidden nodes and various output functions were tested during the model development. Although at present no specific standards exist for the selection of the number of hidden nodes, various guidelines have been proposed in the literature (Rogers and Dowla 1994; Maier and Dandy 1998). Six model architectures were applied for each set of input parameters. The number of applied hidden nodes was $2\,k$, with k varying from 1 to 6. The optimum number of hidden nodes was 8 for the prediction of intestinal enterococci colony forming units and 64 for the prediction of TC colony forming units. The Levenberg–Marquardt optimization method was applied to all models. The Matlab neural network tool box (version 5.3) was used. The ANN model development methodology is similar to one proposed earlier by Mas and Ahlfeld (2007).

The counts of TC and intestinal enterococci per 100 ml in outflow samples collected from 14 April 2005 to 15 September 2006 ranged between 300 and 7,100 and between 300 and 2,010, respectively. The corresponding inflow counts were between 550 and 8,420 and between 360 and 2,130, respectively.

A certain number of relevant inputs should exist to achieve a successful determination of the relationships among the input variables and the model output. When using equations for chemical, biological, or physical processes in a model, the specifications of the processes determine the required input parameters. The selection of inputs is not determined in ANN; therefore, inputs can be selected on the basis of intuitive or empirical understanding of the processes. However, advanced systematic analytical techniques such as PCA or sensitivity analysis can be used when selecting key input parameters (Maier and Dandy 1996; Zhang *et al.* 1998).

When compared with multiple regression analyses, where a p value indicates the significance of a variable and its suitability for inclusion in a model, ANN provide no standard statistical measure to determine the significance of an input variable. Consequently, the input variables (turbidity, pH, conductivity, and dissolved oxygen) selected in this study were chosen on this basis and because of the information gathered from previous literature (Zurada 1992; Sarle 2002; Scholz 2006a).

The data set comprised 60 observed data per parameter per system and was randomly divided into testing, validation, and training data sub-sets. The training set contained 65% of the entries for the entire data set (*i.e.*, 39 observations), whereas the validation and testing sets consisted of 15% (9 observations) and 20% (12 observations) of the entire data set, respectively. Figure 4.19 schemati-

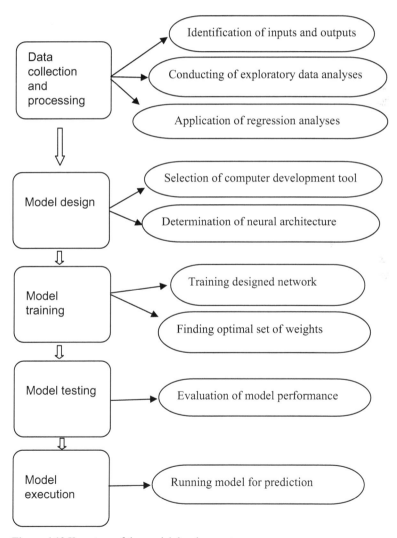

Figure 4.19 Key steps of the model development process

cally shows a series of steps conducted during the model development process (Hamed *et al.* 2004).

4.3.3   Results and Discussion

4.3.3.1   Inflow and Outflow Water Quality

Particularly during warmer seasons, values for 5 d at 20°C BOD (nitrification inhibitor applied), SS, ortho-phosphate–phosphorus, and nitrate–nitrogen were above commonly accepted water quality standard thresholds (25, 35, 2, and 15 mg/l, respectively) for secondary wastewater treatment (EEC 1991). This is partly due to the inflow water quality being representative of the 'worst-case scenario' and the lack of precipitation between 24 March and 13 September 2006, leading to more concentrated inflow water.

Considerable improvements in the quality of the outflow have been observed, particularly when compared to the inflow values. Great improvements were noted predominantly during cold periods for variables such as SS, BOD, and turbidity, whereas most outflow values were considerably below water quality treatment standard thresholds (EEC 1991).

Most European legislation sets a mandatory water quality standard requiring that TC and fecal *Streptococci* should not exceed, respectively, 10,000 CFU per 100 ml and 2,000 CFU per 100 ml for 95% of the water samples (Scholz 2006a). Findings indicate that microbial pollutants very rarely exceed the guideline values even for worst-case simulations, and that they are removed successfully during treatment. The risk of microbial regrowth within detention tanks seems to be low (Kazemi Yazdi and Scholz 2010).

4.3.3.2   Multiple Linear Regression Analyses

The variables BOD and SS can be predicted by applying a multiple linear regression analysis covering 18 months of experimental data. Electrical conductivity, turbidity, pH, ortho-phosphate–phosphorus, nitrate–nitrogen, and ammonia–nitrogen were selected for the prediction because the determination of these variables is less costly and time consuming. Furthermore, stepwise regression was also undertaken to help in the selection of the most appropriate variables for prediction.

TCS and intestinal enterococci colony forming units did not exhibit a significant correlation ($p < 0.05$) with any of the proposed predictors. It was therefore not possible to recommend a successful linear regression model for the microbiological variables obtained in this experiment.

The application of multiple linear regression analyses for the prediction of BOD was relatively successful when applied to samples from the inflow and to systems 1 and 2. This has been attributed to a high correlation between BOD and most of the selected predictors. Moreover, as there has been no strong correlation between BOD and other key water quality variables for system 5, a multiple regression analysis was not performed.

Standard errors of the estimates for SS were higher than the corresponding ones for the BOD. The coefficients of determination (R^2) are relatively high for all systems with the exception of system 5. However, multiple regression analysis is not successful in predicting SS, if a considerable number of outliers are part of the corresponding data set.

4.3.3.3 Analyses of Variance

A one-way ANOVA was conducted to test if the systems performed similarly concerning storm water treatment. The outcome of this analysis allows the design engineer to opt for a system that performs well and is cost-effective. For example, if there is no significant difference between the performances of two different systems for the most important key variables, the designer would be well advised to choose the less costly option.

There were significant differences in treatment performances concerning BOD, ammonia–nitrogen, TC, SS, and intestinal enterococci. Furthermore, the ANOVA indicated that there were significant ($p < 0.05$) differences between some of the water quality parameters in the inflow and outflow for each system. Significant differences with respect to system 1 were found for total dissolved solids, turbidity, electrical conductivity, dissolved oxygen, ortho-phosphate–phosphorus, nitrate–nitrogen, intestinal enterococci counts, and TC counts. For system 2, there were significant differences for turbidity, ortho-phosphate–phosphorus, and nitrate–nitrogen. Concerning system 3, the ANOVA showed significant differences for BOD, total dissolved solids, dissolved oxygen, ortho-phosphate–phosphorus, ammonia–nitrogen, intestinal enterococci, and TC. The results from system 4 indicated significant differences for SS, electric conductivity, ortho-phosphate–phosphorus, nitrate–nitrogen, intestinal enterococci counts, and TC counts. Finally, an ANOVA for system 5 detected significant differences for turbidity, dissolved oxygen, ortho-phosphate–phosphorus, nitrate–nitrogen, intestinal enterococci counts, and TC counts.

The ANOVA showed significant differences between different experimental system performances in treating concentrated road runoff. Systems containing turf showed better BOD and SS removal performances in comparison to less complex systems without turf. However, the assessment was unclear with respect to microbiological indicator variables and the presence or absence of sand mixed with bark.

4.3.3.4 Artificial Neural Network Modeling

The prediction of potentially pathogenic microbiological indicator organisms is important because most of them can only be determined after days of analysis, which makes the information less relevant for real-time system and public health control. The R^2 values for predicting TC counts for the inflow and outflows of systems 1 to 5 were 0.89, 0.94, 0.91, 0.98, 0.59, and 0.95, respectively. The corresponding R^2 values for predicting intestinal enterococci counts were 0.80, 0.63, 0.78, 0.73, 0.71, and 0.15, respectively. It follows that the models were able to successfully predict the TC and intestinal enterococci colony forming unit counts with an exception for the prediction of intestinal enterococci in system 5. Most researchers would consider R^2 values above approx. 0.7 as high for biological data obtained from complex systems (Scholz 2006a; Kazemi Yazdi and Scholz 2010).

The ANN successfully predicted TC and intestinal enterococci counts for the inflow water. The models were very successful in predicting TC counts for all systems except for system 4. Concerning intestinal enterococci counts, the models were relatively successful.

When predicting TC counts with the ANN models for the inflow and for systems 1, 2, 3, and 5, one can undertake predictions confidently resulting in mean squared errors close to zero. In the case of intestinal enterococci counts, the inflow and systems 2, 3, and 4 had similar R^2 values.

The model could be applied outside the experimental set-up for similar problems. The main demand is that the boundary conditions should be comparable. Otherwise, the model would require retraining.

4.3.4 Conclusions

The results of this study show that the ANN models developed for the prediction of TC counts and intestinal enterococci counts perform relatively well. However, the relatively low R^2 values reported for some systems, and more specifically for predicting intestinal entercocci counts in the densely planted system 5, indicate the difficulty in identifying the necessary explanatory variables to characterize a large percentage of the variability observed in the microbial data set. In cases where the water quality standards are observed for TC and intestinal enterococci counts, ANN provide a good modeling technique to predict a potential violation.

Systems containing turf showed better BOD and SS removal performances in comparison to less complex systems without turf. Multiple regression analyses indicated a relatively successful prediction of the BOD, but unsuccessful predictions of both TC and intestinal enterococci counts. However, ANN models predicted both TC and intestinal enterococci counts relatively well.

The neural networks successfully predicted TC and intestinal enterococci counts for most systems. Predictions resulted in mean squared errors that were close to zero.

4.4 Integration of Trees into Drainage Planning

4.4.1 Introduction

4.4.1.1 Background

Urban trees are traditionally an important design element for town planners. Trees and, to a lesser extent, large shrubs can be found at the following urban locations: market places, squares, private English squares (typical in the UK), private yards, near roads (*e.g.*, Kurfürstendamm Allee in Berlin, Germany), rivers and canals (*e.g.*, Grachten in The Netherlands and Germany), private and public gardens, parks (Figure 4.20), car parks (Figure 4.21), beer gardens, tree gardens (usually symmetric layout of trees), garden towns (*e.g.*, Margarethenhöhe in Essen, Germany), terraces (*e.g.*, London terraces), playground, sport grounds (*e.g.*, Olympiapark in Munich, Germany), Tanzlinde (*Tilia* tree around which people dance; typical in central Europe), historical fortification walls (*e.g.*, Naarden in The Netherlands), and graveyards. Color photographs for most of the above examples are shown in Mader and Neubert-Mader (2004).

Urban trees and shrubs are known to have many advantages including those related to noise reduction, air purification, habitat enhancement, and a positive con-

Figure 4.20 Missed opportunity: an urban tree in an Edinburgh park. Why not plant more trees and integrate them within a local sustainable drainage system?

tribution to human psychological well-being (Ellis *et al.* 2006; Everding and Jones 2006; Kuchelmeister and Braatz 1993). Broadleaf deciduous trees are also good phytoremediators for polluted urban air because they have the advantages of higher biomass and faster growth in comparison with woody plants, for example (Takahashi *et al.* 2005). Furthermore, the presence of trees and large shrubs in neighborhoods near retail land use is also beneficial for community satisfaction (Ellis *et al.* 2006). Finally, trees can also improve water management in agriculture by taking up water unused by crops (Oliver *et al.* 2005).

Considerable work has already been done in some of these research areas, but remarkably little is known about the role of trees within the urban hydrological cycle. Although drainage engineers in countries such as the UK are now familiar with the use of sustainable urban drainage systems (SUDS) for storm water management (Stovin *et al.* 2005), it appears that little consideration has been given to the specific role that is (or could be) played by trees and large shrubs.

Legislation in Scotland, for example, actively promotes the use of vegetated SUDS schemes for storm water management in new developments. A number of researchers have demonstrated that it is also feasible, and potentially cost-effective, to introduce SUDS into established urbanized areas affected by insufficient storm water drainage and insufficient management (Stovin *et al.* 2005). This strategy could be defined as SUDS retrofitting (Scholz 2006b).

However, research in the USA, particularly by American Forests (2005), suggests that the land use by urban trees can be directly equated to storm water volumes and, therefore, to the costs of providing engineered structures for storm water management. However, other studies in the UK (*e.g.*, a case study in Sheffield) showed that there is a considerable difference between the percentage of tree cover in the UK compared with some of the cover data cited in the literature published by American Forests (Stovin *et al.* 2005).

Figure 4.21 Example of an experimental belowground detention tank (below the foot path) receiving runoff partially treated with willows

This reflects the higher population densities associated with typical UK urban areas compared with those of cities in the USA (Stovin *et al.* 2005). However, it is important to highlight that the case study in Sheffield was only undertaken for areas with a low tree area cover (<15%) and a high housing density (Stovin *et al.* 2005).

4.4.1.2   Traditional and Sustainable Urban Drainage

Urban drainage in recent decades has focused primarily on reducing the risk of flooding in urban areas and on the associated public health and environmental pollution risks (Butler and Davis 2000). Therefore, urban drainage systems have been concerned mainly with the efficient removal of precipitation, the transport of wastewater to treatment works ('end of pipe' control), and subsequent discharge into receiving waters (*e.g.*, rivers, lakes, and coastal waters).

Combined sewer systems, as typically found in the UK, suffer occasionally from overloading and will subsequently discharge untreated wastewater during times of high rainfall. The environmental impact of storm discharges has been a concern in recent decades. The enhanced storage of storm water is a traditional solution to such problems, but the development of SUDS provides a new option (Struve *et al.* 2002). Most SUDS serve the main purposes of reducing total runoff and therefore its impact on receiving waters, and of reducing peak runoff, which reduces overloading of the sewer system and, therefore, also the frequency of storm discharges.

SUDS include specific types of permeable pavements with enhanced storage or infiltration properties, unpaved areas set aside as natural detention areas such as ponds, and constructed treatment wetlands. Therefore, the provision of storm water storage through such schemes is also directly related to ecological and recreational aspects of integrated catchment management.

It is to be hoped that trees and large shrubs could be integrated either into traditional or sustainable drainage decision making tools (Blanpain *et al.* 2004; Scholz 2006a, b). This section should help to increase the confidence of civil engineering practitioners and hydrological modelers in such options by providing case study site and corresponding calculation examples.

4.4.1.3   Aim and Objectives

The aim of this study is to assess the possibility of integrating trees and large shrubs into water resource management plans for urban areas in Glasgow and Edinburgh. The key objectives are as follows:

- to review the impact of trees on urban hydrology;
- to assess tree integration into traditional and sustainable drainage;
- to assess rainfall and land use characteristics of potential SUDS sites;
- to apply the SUDS Decision Support Model for sites with high proportions of trees and large shrubs;

- to estimate precipitation interception rates for SUDS sites with ≥15% tree area cover; and
- to estimate capital cost savings, if the presence of trees and large shrubs is taken into account in the design of a SUDS.

4.4.2 Methodology

Edinburgh and Glasgow were chosen as case study cities, because they are different in their precipitation patterns (Met Office 2005) and include a large number of distinct residential and park areas with different characteristics that are in their totality typical of and, therefore, representative of most UK cities and urban developments in North America and northern Europe. Glasgow is known to have more precipitation compared with Edinburgh. The mean daily rainfall depths for sites in Glasgow and Edinburgh, and the corresponding tree interceptions, have been calculated (Met Office 2005; Xiao 1998). Potential SUDS construction sites were determined with a methodology (SUDS Decision Support Model) explained in detail elsewhere (Scholz 2006b).

The first part of the study comprised the assessment of areas with similar characteristics for 79 potential SUDS construction sites in Glasgow. However, only ten areas were associated with tree cover areas of ≥15%. The study focused subsequently on 103 potential SUDS construction sites in Edinburgh with different housing densities and tree cover area sizes. However, only 12 areas in Edinburgh were associated with ≥15% tree cover area (mostly parks).

The tree cover area for each site was assessed with aerial photographs (Pauleit and Duhme 2000; Sekliziotis 1980). This was followed by a site survey in 2004 or 2005 to confirm or amend the land use calculations. Trees and large shrubs with a canopy of >2 m were included. For the benefit of this section, large shrubs, which are similar in size and hydrological function to small trees, were also included in the tree survey.

To assess if the integration of trees and large shrubs into a SUDS would make a difference in reducing urban runoff, the tree cover area was related to a reduction in pond size. It was assumed that the pond was the size of a square (sometimes a rectangle) and had a mean depth of 2 m and a mean embankment slope ratio (depth to width) of 1 to 3. The minimum and maximum depths were 1 and 3 m, respectively. The slope ratio was never flatter than 1 to 4 and never steeper than 1 to 2. Interception estimations are usually based on either a 20 or 25 mm d^{-1} rainfall rate depending on the specific site assessed. Moreover, national guidelines in pond design considering the latest research findings were followed if they were suitable for the purpose of the specific site (CIRIA 2005; Scholz et al. 2005; Zheng et al. 2006).

4.4.3   Results and Discussion

4.4.3.1   Lack of Tree Integration into Urban Drainage Systems

SUDS consist of one or more techniques ('treatment train') to manage surface runoff. They are used in conjunction with BMP to prevent flooding of important infrastructure assets and pollution of urban watercourses. There are five general groups of relevant control techniques:

- filter strips;
- belowground detention structures (combined with a vegetated filter, for example, above the detention cells);
- permeable pavement;
- infiltration structures; and
- ponds and constructed wetlands.

The SUDS controls should be located as close as possible to the source of precipitation runoff ('source control') to provide sufficient attenuation. They also provide treatment for surface water using the natural processes of sedimentation, filtration, adsorption, and biological degradation (CIRIA 2005). Most SUDS can be designed to function in urban settings, from hard-surfaced areas to soft-landscaped features. The variety of available design options allows planners and designers to consider local land use, future land management plans, and the wishes of the local population. However, with the exception of some bioretention systems, trees, in contrast to grasses and aquatic plants (predominantly macrophytes), do not feature in the current SUDS philosophy (Butler and Davis 2000; CIRIA 2005; Scholz 2006b).

Trees can serve an important role in storm water mitigation, attenuation, treatment, and infiltration. In terms of the actual mechanical processes, the initial role of trees is to delay the amount of water that reaches the ground below, with rain collected on trees and leaves before continuing its journey downwards. Some of the rainfall will eventually evaporate and some will infiltrate via an infiltration trench, for example, into the ground.

4.4.3.2   Rainfall and Land Use Characteristics

There is often a common general misconception that the whole of Scotland experiences very high rainfall. In fact, rainfall in Scotland varies widely, with a distribution closely related to the topography, ranging from >3000 mm/a in the western highlands and Glasgow to <800 mm/a on the east coast and in Edinburgh. Moreover, rainfall in cold climates also includes large proportions of snow, which subsequently melts, and is frequently expressed as rainfall as well (Met Office 2005).

The temperature profiles are different in both cities throughout the year. For example, the winters in Edinburgh are colder and drier than in Glasgow (Met Office 2005). This results in considerably more high-peak runoff in Glasgow vs. Edinburgh.

Glasgow is characterized by >70 parks and relatively more green space per inhabitant than any other large city in Europe (Glasgow Council 2005). It follows that there is theoretically sufficient space for SUDS and the local integration of trees into the urban water management plan. Nevertheless, Edinburgh is also known for having plenty and somewhat famous green spaces and parks, which are usually located close to the city center (*e.g.*, Holyrood Park with Arthur's Seat and The Meadows).

However, sites with high tree cover areas are usually clustered in urban areas. For example, 50% of the sites with ≥15% tree cover area in Glasgow and Edinburgh have the same first part of the UK postcode (G20 and EH10, respectively). It follows that the integration of existing mature trees into SUDS is unlikely to happen throughout a city.

4.4.3.3 Rainfall Interception

For the purpose of this section, the mean rainfall is defined as the mean monthly precipitation including rain, snow, sleet, and hail. Intense runoff is created during heavy rainfall (*i.e.*, storms) or during snowmelt events. The level of interception is influenced by the following:

- tree group and species (*e.g.*, deciduous, broadleaf evergreen, and conifer-type trees);
- tree architecture (*e.g.*, tree size, number of leaves, and arrangement of leaves and branches);
- rainfall event characteristics (*e.g.*, intensity, duration, and runoff hydrograph); and
- weather (*e.g.*, temperature, relative humidity, net solar radiation, and wind speed).

During a rainfall event, precipitation is either intercepted by leaves, branches, and trunks or it falls directly through the trees to the ground. Intercepted water is stored temporarily, for example, on leaf and bark surfaces. It eventually drips from leaf surfaces and flows down stems and trunk surfaces to the ground, or it evaporates.

Furthermore, healthy trees draw moisture from the soil and ground surface, thereby increasing the long-term soil–water storage potential (CUFR 2005). Tree growth and subsequent decomposition increase the soil water-holding capacity and rate of soil infiltration by rainfall and therefore reduce overland flow. Furthermore, tree canopies reduce soil erosion by diminishing the erosive impact of raindrops on barren surfaces.

Precipitation such as rain and snow often passes through a canopy of vegetation (predominantly trees) before it becomes part of a SUDS. The volume of water retained by this canopy is termed interception. The median vegetation interception, based on several research studies, has been calculated as 13% for deciduous forests and 28% for coniferous woodland (Dunne and Leopold 1978).

Rainfall interception can be very high. For example, a typical medium-size tree can intercept as much as 9,000 l of rainfall. If the tree canopy spans $9\,m^2$, for example, then this represents a rainfall depth of $1\,m/a$ (CUFR 2005). Given that many areas in the UK experience <1 m of rainfall per year, the significance of this figure should not be underestimated. However, comparable and sound UK data are virtually absent. Therefore, it is interesting to consider the influence of urban trees on runoff characteristics. However, the numerical simulation model to estimate annual rainfall interception proposed by Xiao et al. (1998) needs to be adapted to regional conditions for modeling of specific hydrological processes.

The mix of tree species and their corresponding size influence interception. In regions such as Edinburgh and Glasgow, where most precipitation occurs in winter, evergreen trees play the most important role in interception. Trees with evergreen foliage contribute to greater interception than deciduous trees. Some conifers intercept more rainfall than similar sized deciduous trees. In climates with relatively high summer precipitation, deciduous trees such as those in Glasgow, for example, make a substantial contribution to rainfall interception. Typical trees identified in Glasgow and Edinburgh are broadleaf trees; e.g., alder, oak, wych elm, ash, and birch. Planting more trees and improving the health of existing trees is an important strategy, one that helps to reduce the volume of storm water runoff (Geiger 2003). This should be undertaken as part of a SUDS management strategy.

4.4.3.4 Application of the Sustainable Urban Drainage System Decision Support Model

Site characteristics for potential SUDS sites with ≥15% tree cover area are summarized in Table 4.1. Corresponding sites in Edinburgh are less expensive in comparison to Glasgow, because more of these sites are owned by the city council. Edinburgh is dominated by retrofitting of SUDS on council-owned sites, while Glasgow has more sites associated with development and regeneration as well as with recreational sites. It follows that Glasgow has a high proportion of future industrial, institutional, and commercial roof and car park runoff to be treated by SUDS in comparison to Edinburgh (Table 4.1).

Sites with a high proportion of trees in comparison to other sites are less attractive for development (13 to 27%) but more attractive for retrofitting (32 to 14%) of a SUDS, if compared to the overall database for both cities. This is beneficial for the protection of existing mature trees. The groundwater level for sites with a lot of trees is also relatively low (i.e., close to the ground surface) in comparison to the overall data set (91 to 82%), which is beneficial for enhancing the infiltration performance. Finally, the potential for a high ecological impact is also greater for sites with a lot of trees (82 to 25%) (Table 4.2 and Scholz 2006b).

Table 4.2 Proportions of selected site characteristics for potential sustainable urban drainage system (SUDS) sites with ≥15% tree cover in Glasgow and Edinburgh

SUDS technique	Glasgow	Edinburgh	Both cities
Council ownership of site (%)	80	100	91
Relative land cost estimation (0 = inexpensive; 100 = expensive)	76	48	61
Development site (%)	20	8	13
Regeneration site (%)	50	33	41
Recreation site (%)	30	0	14
SUDS retrofitting only (%)	0	59	32
Proportion of site area suitable for SUDS (%)	90	80	85
High groundwater table (%)	10	8	9
Mean site slope (%)	7	4	5
Current impermeable area (%)	22	22	22
Future impermeable area after SUDS implementation (%)	50	55	53
Road runoff in future (%)	100	100	100
Domestic roof runoff in future (%)	80	92	87
Institutional roof runoff in future (%)	50	25	36
Industrial roof runoff in future (%)	40	8	23
Commercial roof runoff in future (%)	20	0	9
Car park runoff in future (%)	30	17	23
Drainage to combined sewer possible (%)	100	92	96
Drainage to local watercourse possible (%)	90	50	68
Potential for high ecological impact after SUDS implementation (%)	60	100	82

Table 4.3 Proportions of sustainable urban drainage system (SUDS) techniques (%) for sites with ≥15% tree cover (in comparison to all sites) in Glasgow and Edinburgh, based on the SUDS Decision Support Model (Scholz 2006b)

SUDS technique	Glasgow		Edinburgh		Both cities	
	≥15%	All	≥15%	All	≥15%	All
Wetland	13	3	5	0	8	1
Attenuation or detention pond	9	17	15	14	13	15
Infiltration pond or basin	9	10	14	19	13	11
Swale	18	6	9	10	12	8
Filter strip	4	7	13	6	10	6
Soakaway	4	8	13	9	10	9
Infiltration trench	9	6	11	8	10	7
Permeable pavement	4	16	5	16	5	16
Belowground storage	13	9	4	6	7	7
Water playground	4	8	7	9	6	9
Green roof	13	10	4	11	6	11

Table 4.3 shows the results of the SUDS Decision Support Model application for sites with ≥15% tree area cover. Swales combined with ponds or wetlands were recommended for approx. 46% of all corresponding sites. Attenuation, detention, or infiltration ponds accounted for 26% of all recommended SUDS options. This was the justification for using ponds as an example SUDS technique where the dimensions could be reduced if trees and large shrubs were integrated into the SUDS design. Moreover, it is remarkable that wetlands (13 to 3%) and swales (18 to 6%) were particularly recommended by the model for sites with a high proportion of trees in Glasgow.

4.4.3.5 Design Recommendations

The interception rate of trees is directly linked to the attenuation and reduction of storm runoff. For example, Equation 4.4 shows the interception uptakes of urban trees (mm), y, for different rainfall events (mm), x, in Glasgow and Edinburgh. In comparison, Equation 4.5 shows the interception uptakes of urban trees (m^3), y, for tree cover areas (m^2), x, assuming a 10 mm rainfall event for SUDS sites in Glasgow and Edinburgh.

$$y = -21.81 \ \ln(x) + 109.92; \ R^2 = 0.97, \tag{4.4}$$

$$y = 0.01 \times + 0.03; \ R^2 = 1.00. \tag{4.5}$$

The integration of urban trees into SUDS could save land and construction costs. Trees are able to lower the costs for swale-like and other linear conveyance structures, infiltration trenches, bioretention areas, ponds, and wetlands. Without trees to attenuate, treat, and reduce the runoff, additional storage and treatment facilities, which are very expensive for developers and local taxpayers alike, would need to be constructed.

The potential benefit of urban trees is predominantly a crude function of the tree cover area. The survey in Edinburgh and Glasgow showed that a tree cover area of only ≥15% justifies integrating trees into SUDS, if savings of ≥10% could be achieved. For ponds, for example, trees are predominantly effective in intercepting rainfall during small storm events. Therefore the role of urban trees may be more significant in terms of urban water quality and biodiversity enhancement than in reducing serious flooding events (*e.g.*, recent Boscastle, England, flood).

Figures 4.22 and 4.23 show how trees could reduce the size of SUDS ponds and subsequently reduce land and construction costs in Glasgow and Edinburgh, respectively. Only areas with ≥15% tree area cover have been included.

This analysis is based on the estimation that the enhanced interception capability of trees as part of a SUDS structure incorporating a pond is approx. 40% for a mean rainfall event for all sites researched in Glasgow and Edinburgh. Calculations are backed up by case studies in Glasgow (Ruchill Hospital) and Edinburgh

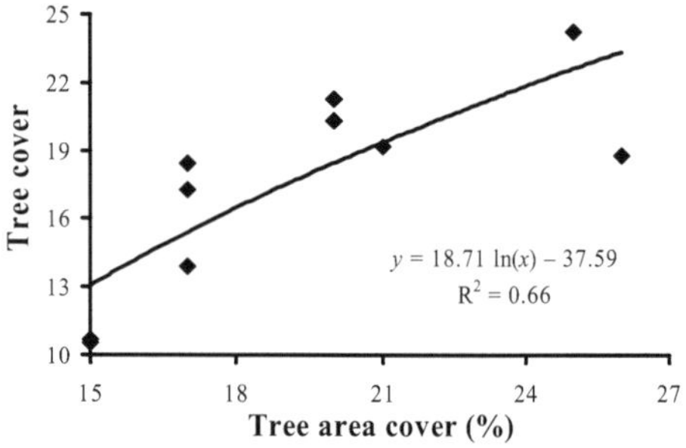

Figure 4.22 Relationship between tree cover (%), x, and corresponding reduction of pond size (m²), y, in Glasgow, if trees are integrated into sustainable urban drainage system design of relevant sites with ≥15% tree cover

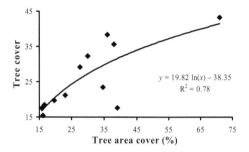

Figure 4.23 Relationship between tree cover (%), x, and corresponding reduction of pond size (m²), y, in Edinburgh, if trees are integrated into sustainable urban drainage system design of relevant sites with ≥15% tree cover area

(The King's Buildings) as described elsewhere (Scholz *et al.* 2005; Zheng *et al.* 2006). For an increase in tree cover area from approx. 25 to 50%, the results indicate that the corresponding mean potential reduction in annual runoff ranges between 10 and 20%. It follows that a retention pond area in Edinburgh, for example, could be reduced by approx. 10% (Figure 4.23). Similar findings have been reported by the Canada Mortgage and Housing Cooperation (CMHC 2005).

Table 4.4 Relationship between tree cover (%) and the corresponding reduction of pond size (m²) in Glasgow if trees are integrated into the sustainable urban drainage system design of relevant large sites with <15% tree cover

Site area (m²)	Tree area cover (%)	Saved pond area (%)
252,000	7	5.2
215,600	5	4.2
208,000	13	10.3
191,540	12	10.0
136,000	4	4.0
95,500	6	6.4
92,400	9	9.4
58,800	1	0.8
45,890	3	3.4
38,270	2	2.7

Table 4.5 Relationship between tree cover (%) and the corresponding reduction of pond size (m^2) in Edinburgh, if trees are integrated into the sustainable urban drainage system design of relevant large sites with <15% tree cover

Site area (m^2)	Tree area cover (%)	Saved pond area (%)
234,375	9	6.0
156,250	4	3.6
125,000	1	0.6
62,500	14	14.3
62,500	5	5.5
62,500	2	2.3
47,500	12	12.9
31,250	7	7.9
15,625	11	12.1
15,625	3	3.8
11,875	2	3.0

Finally, Tables 4.4 and 4.5 summarize the tree survey findings for relevant large sites with <15% tree cover for Glasgow and Edinburgh, respectively. Almost 200 sites were surveyed, but only large sites are shown. It can be seen that the saved pond areas are usually small for sites with low proportions of tree cover areas and that there is no obvious relationship between the site area size and the corresponding area covered by trees.

4.4.4 Conclusions

Urban trees play an important role in the urban hydrological cycle, yet little consideration has been given in the UK to their integration into urban drainage strategies. Tree planting can be justified on the basis of financial benefits associated with their storm water attenuation and reduction function alone. However, trees also provide other benefits for the urban environment (*e.g.*, increased biodiversity).

Guidance on estimating interception with rainfall or tree cover was given. The study also showed that for areas with ≥15% tree cover, the area for ponds as part of a SUDS treatment train, for example, could be reduced by integrating trees into the design of the SUDS such as infiltration trenches and bioretention areas, which could save ≥10% of the capital costs. Swales combined with ponds were the most suitable SUDS options for most sites.

However, not every tree is convenient for each area. Green broadleaf trees are usually preferable. Furthermore, the exact site location of already existing trees in terms of SUDS retrofitting is an important factor for detailed SUDS design.

References

American Forests (2005) Various documents. www.americanforests.org. Accessed 20 Dec 2009
Anderson DM, Gilbert PM, Burkholder JM (2002) Harmful algal blooms and eutrophication: nutrient sources, composition, and consequences. Estuaries 25:704–726
ATV–DVWK (2001) Hochwasserrückhaltebecken – Probleme und Anforderungen aus wasserwirtschaftlicher und ökologischer Sicht. Deutsche Vereinigung für Wasserwirtschaft, Abwaser und Abfall eV (ATV–DVWK), Gesellschaft zur Förderung der Abwassertechnik eV, Hennef, Germany
Bah AR, Kravchuk O, Kirchhof G (2009) Sensitivity of drainage to rainfall, vegetation and soil characteristics. Comput Electron Agric 68:1–8
Bayley SE, Guimond JK (2008) Effects of river connectivity on marsh vegetation community structure and species richness in montane floodplain wetlands in Jasper National Park, Alberta, Canada. Ecoscience 15:377–388
Blanpain O, Karnib A, Al-Hajjar J (2004) A decision making tool providing aid in choosing a storm drainage network solution: description and analysis. Urban Wat J 2004:217–226
Brion GM, Lingireddy S (2003) Artificial neural network modeling: a summary of successful applications relative to microbial water quality. Wat Sci Technol 47:235–240
Bronstert A, Baardossy A, Bismuth C, Buiteveld H, Disse M, Engel H, Fritsch U, Hundecha Y, Lammersen R, Niehoff D, Ritter N (2007) Multi-scale modeling of land-use change and river training effects on floods in the Rhine basin. River Res Appl 23:1102–1125
Brown RB (2003) Soil texture. University of Florida, Institute of Food and Agricultural Sciences Extension. http://edis.ifas.ufl.edu/SS169. Accessed 30 Dec 2009
Bulc T, Slak AS (2003) Performance of constructed wetland for highway runoff treatment. Wat Sci Technol 48:315–322
Butler D, Davis JW (2000) Urban drainage. E&FN Spon, London, UK
Butler D, Parkinson J (1997) Towards sustainable urban drainage. Wat Sci Technol 35:53–63
Butler D, Friedler E, Gatt K (1995) Characterising the quantity and quality of domestic wastewater inflows. Wat Sci Technol 31:13–24
Casali J, Gimenez R, De Santisteban L, Alvarez-Mozos J, Mena J, Del Valle de Lersundi J (2009) Determination of long-term erosion rates in vineyards of Navarre (Spain) using botanical benchmarks. Catena 78:12–19
Chung EG, Bombardelli FA, Geoffrey Schladow S (2009) Modeling linkages between sediment resuspension and water quality in a shallow, eutrophic, wind-exposed lake. Ecol Model 220:1251–1265
CEH (1999) Flood estimation handbook. Institute of Hydrology, CEH Wallingford, England, UK. http://www.nwl.ac.uk/ih/feh/html/handbook.html. Accessed 10 Jan 2010
CIRIA (2004) The SUDS manual (C697). Construction Industry Research and Information Association (CIRIA), London, UK
CIRIA (2005) Sustainable drainage systems: promoting good practice. Construction Industry Research and Information Association (CIRIA). www.ciria.org/suds. Accessed 20 Dec 2009
CMHC (2005) Various documents. www.cmhc-schl.gc.ca/en/imquaf/himu/wacon/wacon_027.cfm, example document. Accessed 25 Dec 2009
Colin F, Puech C, de Marsily G (2000) Relations between triazine flux, catchment topography and distance between maize fields and the drainage network. J Hydrol 236:139–152
Collins AL, Walling DE (2007) The storage and provenance of fine sediment on the channel bed of two contrasting lowland permeable catchments, UK. Riv Res Appl 23:429–450
CUFR (2005) Trees reduce storm water runoff. Center for Urban Forest Research (CUFR). www.cufr.ucdavis.edu/research/water.asp. Accessed 20 Dec 2009
Dawson JJC, Soulsby C, Tetzlaff D, Hrachowitz M, Dunn SM, Malcolm IA (2008) Influence of hydrology and seasonality on DOC exports from three contrasting upland catchments. Biogeochemistry 90:93–113
Dunne T, Leopold LB (1978) Water in environmental planning. Freeman, New York, NY, USA

EDINA (2009) Digimap and Historic Digimap Collections. http//edina.ed.ac.uk. Accessed 10 Jan 2010

Edwards AC, Withers PJA (2008) Transport and delivery of suspended solids, nitrogen and phosphorus from various sources to freshwaters in the UK. J Hydrol 350:144–153

EEC (1991) EEC urban waste water treatment. Directive 91/271/EEC 1991, European Economic Community, Brussels, Belgium

Ellis B (1991) Urban runoff quality in the UK – problems, prospects and procedures. Appl Geogr 11:187–200

Ellis CD, Lee S-W, Kweon B-S (2006) Retail land use, neighbourhood satisfaction and the urban forest: an investigation into the moderating and mediating effects of trees and shrubs. Landsc Urban Plann 74:70–78

Environment Agency (2009) http://www.environment-agency.gov.uk/research/planning/33240.aspx. Accessed 1 Aug 2009

EU (2007) The quality of drinking water in the European Union synthesis report on the quality of drinking water in the European Union period 2002–2004. Directives 80/778/EEC and 98/83/EC. http://circa.europa.eu/Public/irc/env/drinking_water_rev/library?l=/drinking_synthesis/report_2002-2004pdf/_EN_1.0_&a=d. Accessed 2 Dec 2009

Everding SE, Jones DN (2006) Communal roosting in a suburban population of Torresian crows (*Corvus orru*). Landsc Urban Plann 74:21–33

Geiger JR (2003) Is all your rain going down the drain? Research summary in Center for Urban Forest Research, Pacific Southwest Research Station, USDA Forest Service, p 4

Gerasimo IP (1969) Revision of genetic principles of Dokuchayevs soil science in new American classification of soils and in works on compilation of world soil map. Soviet Soil Sci – USSR 5:511

Gervin L, Brix H (2001) Removal of nutrients from combined sewer overflows and lake water in a vertical-flow constructed wetland system. Wat Sci Technol 44:171–176

Glasgow Council (2005) City of parks and gardens. www.glasgow.gov.uk/en/AboutGlasgow/Touristattractions/cityofparksandgardens.htm. Accessed 20 Dec 2009

GPS (2009) Magellan Geographical Positioning System (GPS) receivers. http://www.gps-maps.us/magellan-gps-receivers.shtml. Accessed 20 Nov 2009

Hamed MM, Khalafallah MG, Hassanian EH (2004) Prediction of wastewater treatment plant performance using artificial neural networks. Environ Model Softw 19:919–928

He LM, He ZL (2008) Water quality prediction of marine recreational beaches receiving watershed baseflow and stormwater runoff in southern California, USA. Wat Res 42:2563–2573

Heal KAV (2001) Manganese and land-use in upland catchments in Scotland. Sci Total Environ 265:169–179

House A (2009) River basin district. http://www.euwfd.com/html/river_basin_districts.html. Accessed 10 Aug 2009

IPCC 2007. Contribution of Working Group II to the Fourth Assessment Report of the Intergovernmental Panel on Climate Change. In: Parry ML, Canziani OF, Palutikof JP, van der Linden PJ, Hanson CE (eds) Climate Change 2007: impacts, adaptation and vulnerability. Cambridge University Press, Cambridge, UK

IUCN (2000) Vision for water and nature. A world strategy for conservation and sustainable management of water resources in the 21st century. Compilation of all project documents. Cambridge, UK

Iyengar SS, Kashyap RL (1989) Autonomous intelligent machines. Computer June:14–15

Jeng HC, England AJ, Bradford HB (2005) Indicator organisms associated with storm water suspended particles and estuarine sediment. J Environ Sci Health 40:779–791

Johnson B, Balserak P, Beaulieu S, Cuthbertson B, Stewart R, Truesdale R, Whitmore R, Young J (2003) Industrial surface impoundments: environmental settings, release and exposure potential and risk characterization. Sci Total Environ 317(1–3): 1–22

Kaufman L, Rousseeuw, PJ (1990) Finding groups in data – an introduction to cluster analysis. Wiley, Hoboken, NJ, USA

Kay D, Wyer M, Crowther J, Stapleton C, Bradford M, McDonald A, Greaves J, Francis C, Watkins J (2005) Predicting faecal indicator fluxes using digital land use data in the UK's sentinel Water Framework Directive catchment: the Ribble study. Wat Res 39:3967–3981

Kazemi Yazdi S, Scholz M (2010) Assessing storm water detention systems treating road runoff with an artificial neural network. Wat Air Soil Pollut 206:35–47

Kendrick M (1988) The Thames barrier. Landsc Urban Plann 16:57–68

Kuchelmeister G, Braatz S (1993) Urban forestry revisited. Unasylva J Forestry 173:13–18

Kuhn G, Diekmann B (2003) Data report: bulk sediment composition, grain-size, clay, and silt mineralogy of Pleistocene sediments from ODP Leg 177 Sites 1089 and 1090. Proc ODP Sci Results 177:1–10

Larinier M (2000) Dams and Fish Migration. World Commission on Dams. FAO Fisheries Technical paper, 419:45–89

Lee B-H, Scholz M (2006) A comparative study: prediction of constructed treatment wetland performance with K-nearest neighbours and neural networks. Wat Air Soil Pollut 174:279–301

Loures L, Panagopoulos T (2007) From Derelict Industrial Areas towards Multifunctional Landscapes and Urban Renaissance. WSEAS Trans Environ Develop 3(10):181–188

Mader G, Neubert-Mader L (2004) Bäume – Gestaltungsmittel in Garten, Landschaft und Städtebau. Komet, Cologne, Germany (in German)

Maier HR, Dandy GC (1996) The use of artificial neural networks for the prediction of water quality parameters. Wat Resour Res 32:1013–1022

Maier HR, Dandy GC (1998) The effect of internal parameters and geometry on the performance of back propagation neural networks: an empirical study. Environ Model Softw 13:193–209

Maier HR, Dandy GC (2000) Neural networks for the prediction and forecasting of water resources variables: a review of modeling issues and applications. Environ Model Softw 15:101–124

Mas DML, Ahlfeld DP (2007) Comparing artificial neural networks and regression models for predicting faecal coliform concentrations. Hydrol Sci J 52:713–731

McCorquodale JA, Georgiou I, Carnelos S, Englande AJ (2004) Modeling coliforms in storm water plumes. J Environ Eng Sci 3:419–431

McLemore VT (1988) Chicosa Lake State Park. New Mexico Geol 10(3):62–64

Memon FA, Butler D (2002) Assessment of gully pot management strategies for runoff quality control using a dynamic model. Sci Total Environ 295:115–129

Met Office (2005) Climate. Meteorological Office of the United Kingdom (Met Office). www.metoffice.co.uk. Accessed 20 Dec 2009

Minitab Inc. (2003) Minitab Statistical Software, Release 14 for Windows. State College, PA, USA

Mitchell GN (1991) Water quality issues in the British uplands. Appl Geogr 11(3):201–214

Mohanty S, Scholz M, Slater MJ (2002) Neural network simulation of the chemical oxygen demand reduction in a biological activated carbon filter. J Ch Inst Wat Environ Manag 16:58–64

Mohssen M (2008) An insight into flood frequency for design floods. Flood Recovery Innovat Response 118:155–164

Montaldo N, Mancini M, Rosso R (2004) Flood hydrograph attenuation induced by a reservoir system: analysis with a distributed rainfall-runoff model. Hydrol Process 18(3):543–563

Moss B (1988) Ecology of Freshwaters; Man and Medium, 2nd edn. Blackwell, Oxford

Neelakantan TR, Brion GM, Lingireddy S (2001) Neural network modeling of Cryptosporidium and Giardia concentrations in the Delaware River, USA. Wat Sci Technol 43:125–132

Nisbet TR, Welch D, Doughty R (2002) The role of forest management in controlling diffuse pollution from the afforestation and clear felling of two public water supply catchments in Argyll, West Scotland. For Ecol Manag 158(1–3):141–154

NRC (2004) Indicators for waterborne pathogens. National Research Council of the National Academies (NRC), National Academies Press, Washington, DC

O'Keefe B, D'Arcy BJ, Barbarito B, Clelland B (2005) Urban diffuse sources of faecal indicators. Wat Sci Technol 51:183–190

Olding DD, Steele TS, Nemeth JC (2004) Operational monitoring of urban stormwater management facilities and receiving subwatersheds in Richmond Hill, Ontario. Wat Qual Res J Can
39:392–405

Oliver YM, Lefroy EC, Stirzaker R, Davies CL (2005) Deep-drainage control and yield: the
trade-off between trees and crops in agroforestry systems in the medium to low rainfall areas
of Australia. Austr J Agric Res 56:1011–1026

Pauleit S, Duhme F (2000) Assessing the environmental performance of land cover types for
urban planning, Landsc Urban Plann 52:1–20

Pratap R (2002) Getting Started with MATLAB: A Quick Introduction for Scientists and Engineers. Oxford University Press, Oxford, UK

Ramchunder SJ, Brown LE, Holden J (2009) Environmental effects of drainage, drain-blocking
and prescribed vegetation burning in UK upland peatlands. Prog Phys Geogr 33(1):49–79

Ramos-Scharron CE, MacDonald LH (2007) Measurement and prediction of natural and anthropogenic sediment sources, St John, US Virgin Islands. Catena 71(2):250–266

Rogers LL, Dowla FU (1994) Optimization of groundwater remediation using artificial neural
networks with parallel solute transport modeling. Wat Resour Res 30:457–481

Sandhu N, Finch R (1996) Emulation of DWRDSM using artificial neural networks and estimation of Sacramento River flow from salinity. In: Proceedings of the North American water
and environment conference, ASCE, New York, pp. 4335–4340

Sarle W (2002) The neural network FAQ. ftp://ftp.sas.com/pub/neural/FAQ4.html. Accessed
Nov 2007

Scholz M (2006a) Wetland systems to control urban runoff. Elsevier, Amsterdam, The Netherlands

Scholz M (2006b) Practical sustainable urban drainage system decision support tools. Proc Inst
Civ Eng Eng Sustain 159:117–125

Scholz M (2007a) Classification methodology for sustainable flood retention basins. Landsc
Urban Plann 81(3):246–256

Scholz M (2007b) An expert classification system for sustainable flood retention basins. Civ Eng
Environ Syst 24(3):193–209

Scholz M (2007c) Ecological effects of water retention in the River Rhine Valley: a review
assisting future retention basin classification. Int J Environ Stud 64:171–187

Scholz M, Sadowski AJ (2009) Conceptual classification model for sustainable flood retention
basins. J Environ Manag 90(1):624–633

Scholz M, Morgan R, Picher A (2005) Storm water resources development and management in
Glasgow: two case studies. Int J Environ Stud 62:263–282

Sekliziotis S (1980) A survey of urban open space using colour-infrared aerial photographs.
PhD thesis, University of Aston, Birmingham, UK

SEPA (2007) Scotland's WFD aquatic monitoring strategy, Scottish Environment Protection
Agency (SEPA), Stirling, UK

Schumacher BA (2002) Methods for the determination of total organic carbon (TOC) in soils and
sediments. Ecological Risk Assessment Support Center, Office of Research and Development, US Environmental Protection Agency, Washington, DC, USA

Shepherd TJ, Chenery SRN, Pashley V, Lord RA, Ander LE, Breward N, Hobbs SF, Horstwood
M, Klinck BA, Worrall F (2009) Regional lead isotope study of a polluted river catchment:
River Wear, Northern England, UK. Sci Total Environ 407(17):4882–4893

SPSS (2009) Statistical Package for the Social Sciences (SPSS). Recently re-branded as Pre-
dictive Analytics SoftWare (PASW). http://www.spss.com/uk/statistics/?gclid=CNvH2-
XFuZsCFY4U4wodMkPQEA. Accessed 1 Dec 2009

Stevens CJ, Quinton JN (2009) Diffuse pollution swapping in arable agricultural systems.
Crit Rev Environ Sci Technol 39(6):478–520

Stovin VR, Jorgensen A, Clayden A, Swan A (2005) Urban trees and stormwater management –
an undervalued resource? In: Third National Conference on Sustainable Drainage. Incorporating 27th Meeting of Standing Conference on Stormwater Source Control, 20–21 June 2005,
Coventry University, pp. 53–63

Strategic Alliance for Water Management Actions (2009) SAWA – A Strategic Alliance in the North Sea Region. Strategic Alliance for Water Management Actions (SAWA) project information. http://www.sawa-project.eu/. Accessed 1 Dec 2009

Struve J, Westen S, Millard K, Fortune D (2002) State of the art review. HR Wallingford Report Number EVK1-CT-2001-00090-SR 598

Sutherland WJ (2006) Ecological census techniques – a handbook, 2nd edn. Cambridge University Press, Cambridge, UK

Takahashi M, Higaki A, Nohno M, Kamada M, Okamura Y, Matsui K, Kitani S, Morikawa H (2005) Differential assimilation of nitrogen dioxide by 70 taxa of roadside trees at an urban pollution level. Chemosphere 61:633–639

Tomenko V, Ahmed S, Popov V (2007) Modeling constructed wetland treatment system performance. Ecol Model 205:355–364

UK Groundwater Forum (2009) Groundwater basics. http://www.groundwateruk.org/What-is-Groundwater.aspx. Accessed 10 Aug 2009

Ukonmaanaho L, Nieminen TM, Rausch N, Cheburkin A, Le Roux G, Shotyk W (2006) Recent organic matter accumulation in relation to some climatic factors in ombrotrophic peat bogs near heavy metal emission sources in Finland. Glob Planet Change 53(4):259–268

Watzin MC, McIntosh AW (1999) Aquatic ecosystems in agricultural landscapes: a review of ecological indicators and achievable ecological outcomes. J Soil Wat Conserv 54(4):636–644

Wu JS, Ahlert RC (1978) Assessment of methods for computing storm runoff loads. J Am Wat Resour Assoc 14:429–439

Wu J, Yu SL, Zou R (2006) A water quality based approach for watershed wide BMP strategies. J Am Wat Resour Assoc 42:1193–1204

Xiao Q, McPherson EG, Simpson JR, Ustin SL (1998) Rainfall interception by Sacramento's urban forest. J Arboric 24:235–244

Zhang G, Patuwo BE, Hu MY (1998) Forecasting with artificial neural networks: the state of the art. Int J Forecast 14:35–62

Zhang Y, van Dijk MA, Liu M, Zhu G, Qin B (2009) The contribution of phytoplankton degradation to chromophoric dissolved organic matter (CDOM) in eutrophic shallow lakes: field and experimental evidence. Wat Res 43:4685–4697

Zheng J, Nanbakhsh H, Scholz M (2006) Case study: design and operation of sustainable urban infiltration ponds treating storm runoff. J Urban Plann Develop Am Soc Civ Eng 132:36–41

Zurada JM (1992) Introduction to artificial neural systems. West Publishing, St Paul, London, UK

Chapter 5
Modeling Complex Wetland Systems

Abstract This chapter focuses on the self-organizing map (SOM) model, which was applied to predict outflow nutrient concentrations for ICW treating farmyard runoff. The SOM showed that the outflow ammonia–nitrogen concentrations were strongly correlated with water temperature and salt concentrations, indicating that ammonia–nitrogen removal is effective as low salt concentrations and comparatively high temperatures in ICW. The SRP removal was predominantly affected by salt and dissolved oxygen concentrations. In addition, pH and temperature were weakly correlated with SRP removal, suggesting that SRP was easily removed within ICW if salt concentrations were low and dissolved oxygen, temperature, and pH values were high. The SOM model performed very well in predicting the nutrient concentrations with water quality variables such as temperature, conductivity, and dissolved oxygen, which can be measured cost-effectively. The results indicate that the SOM model was an appropriate approach to monitor wastewater treatment processes in ICW.

5.1 Introduction

Constructed wetlands are often used as artificial wastewater treatment systems usually composed of one or more treatment cells that are planted with aquatic vegetation such as macrophytes (USEPA 2000). They are used to treat many types of wastewater including urban runoff, municipal and industrial wastewaters, agricultural runoff and wastewater, and acid mine drainage (USEPA 2000; Scholz 2006).

Constructed wetlands are usually efficient in reducing COD, BOD, and SS, but the corresponding removal efficiencies for nitrogen and phosphorus are often low (USEPA 2000; Vymazal 2007). Nitrogen and phosphorus are considered to be the most important nutrients causing water pollution. Nitrogen has an intricate biogeochemical cycle with various biotic and abiotic transformations. The important inorganic forms of nitrogen in wetlands are ammonia–nitrogen, nitrate–nitrogen, and nitrite–nitrogen according to Vymazal (2007). Phosphorus occurs predominantly as phosphate in natural waters and wastewater. Phosphates are classified as ortho-phosphate, condensed (pyro-, meta-, and poly-) phosphates, and organically bound phosphate (USEPA 2000).

The treatment mechanisms and processes within constructed wetlands are highly complex and include microbial, biological, physical, and chemical processes that may occur sequentially or simultaneously (Hammer and Bastian 1989; USEPA 2000; Scholz 2006; Vymazal 2007). The processes of nitrogen removal and retention during wastewater treatment in constructed wetlands are manifold and include ammonia volatilization, nitrification, denitrification, nitrogen fixation, plant and microbial uptake, mineralization (ammonification), nitrate ammonification, anaerobic ammonia oxidation, ammonia adsorption and burial (Vymazal 2007). Phosphorus removal and retention mechanisms during wastewater treatment in constructed wetlands include adsorption, desorption, precipitation, plant and microbial uptake, fragmentation, leaching, mineralization, sedimentation (peat accretion), and burial (Hammer and Bastian 1989; Vymazal 2007).

In comparison to conventional constructed wetlands, which cannot remove significant amounts of nitrogen and phosphorus, ICW are a more effective type of constructed wetland suitable for nitrogen and phosphorus removal (USEPA 2000; Harrington and Ryder 2002; Harrington *et al.* 2005). The ICW concept promoted by the ICW Initiative of the Irish National Parks and Wildlife Service has been described in detail by Scholz *et al.* (2007). Integrated constructed wetlands are free surface-flow constructed wetlands, which are based on the holistic use of land to maintain water quality and include elements of good landscape fit, biodiversity, and habitat enhancement in their design (Carroll *et al.* 2005).

Modeling and predicting treatment processes is significant for elucidating the complex nutrient removal mechanisms and assessing the corresponding water treatment potential of ICW. It is necessary to model and predict the nutrient removal processes to optimize the design, operation, management, and water quality monitoring strategy of an ICW.

The SOM is based on an unsupervised neural network algorithm and has been used to analyze, cluster, and model various types of large databases (Kohonen *et al.* 1996; Lee and Scholz 2006). Astel *et al.* (2007) and Scholz (2008) applied SOM models successfully for the classification of large water and environmental data sets. The advantages of the SOM algorithm and its classification and visualization ability have been exploited. The SOM was used as the first abstraction level in clustering. The original data set was represented using a smaller set of prototype vectors, which allowed for the efficient use of clustering algorithms to divide the prototypes into groups, and the 2-D grid allowed for a rough visual presentation and interpretation of the clusters as outlined by Vesanto and Alhoniemi (2000).

The SOM model, which has not been implemented in water treatment process control strategies as often as traditional neural networks, was successfully used for the first time as a prediction tool for heavy metal removal in constructed wetland systems by Lee and Scholz (2006). However, the SOM model has never been applied to model and predict the nitrogen and phosphorus removal efficiencies within constructed wetland systems such as ICW.

The aims of this study were as follows:

1. to assess the farmyard runoff treatment potential in terms of nutrient removal with an ICW;
2. to identify relationships between nutrients and other water quality variables; and
3. to predict the nutrient concentration removal performances with the SOM model using water quality parameters that are more cost-effective, quicker, and easier to measure.

5.2 Methodology and Software

5.2.1 Case Study Sites

The ICW study presented in this section relates to 13 ICW systems, which were constructed to treat farmyard runoff and wastewater within the Anne Valley near Waterford in Ireland. The farmyard runoff and waste entering an ICW typically consists of yard and dairy washings and rainfall on open yard and farmyard roofed areas along with silage (usually only spillages) and manure (occasional droppings) effluents. Construction of the ICW systems began in 2000 and was followed by commissioning in February 2001. Scholz *et al.* (2007) describe these systems and their catchments in detail.

The ICW 3, 9, and 11 were built on dairy farms operated for 50, 55, and 77 cows, respectively. The corresponding wetland sizes were 10,288, 7,964, and 7,676 m². The ICW 9 and 11 had four cells, while ICW 3 had five cells; all wetland cells had a linear sequential arrangement. The mean ICW size was approx. 1.7 times the size of the farmyard areas. The primary vegetation types planted in the ICW systems were emergent species (helophytes). Figures 5.1 and 5.2 show the ICW 11 system in winter.

Figure 5.1 Sedimentation tank of representative integrated constructed wetland system 11 in winter 2006

Figure 5.2 Inlet arrangement of the first cell of the representative integrated constructed wetland system 11 in winter 2006

5.2.2   Data and Variables

The ICW data were collected by monitoring the inflow and outflow water qualities of all 13 ICW systems for more than 6 years (August 2001 to December 2007). However, this section is based on only a fraction of the overall data set to address the corresponding aims. Only data obtained from the representative and typical ICW system sites 3, 9, and 11 (characterized by Scholz *et al.* (2007)) were combined and subsequently used in this section because these systems have linear sequential cell configurations and single influent entry points. In contrast, the other ICW have either multiple influent entry points or parallel treatment cells. All three selected ICW sites are typical FCW (specific application of ICW to treat farmyard runoff), previously defined by Carty *et al.* (2008).

Water samples were analyzed for ammonia–nitrogen, SRP, dissolved oxygen (DO), temperature, pH, chloride, and conductivity according to standard methods (Allen 1974; APHA 1998). Ammonia–nitrogen and chloride were determined using automated colorimetry. Soluble reactive phosphorus was determined as MRP with an auto-analyzer (Method 2540-D; APHA 1998). DO, temperature, pH, and conductivity were measured in the field with portable meters. Scholz *et al.* (2007) provides a detailed description of the water quality analysis.

The inexpensive and easy-to-measure SOM input water quality variables of the outflow were DO (mg/l), temperature (°C), pH (–), chloride (mg/l), and conductivity (µS). The corresponding expensive and time-consuming-to-measure model output parameters were outflow ammonia–nitrogen (mg/l) and SRP (mg/l).

5.2.3   Statistical Analyses

All statistical analyses were performed using the standard software packages Origin 7.0, Matlab 7.0, and Econometrics Views 5.0. Significant differences (usually $p < 0.05$, unless stated otherwise) between data sets are indicated where appropriate.

5.2.4 Self-organizing Map

The SOM is a neural network model and algorithm that implements a characteristic non-linear projection from the high-dimensional space of sensory or other input signals onto a low-dimensional array of neurons and has been widely applied to the visualization of dimensional systems and data mining (Kohonen *et al.* 1996). The SOM is a competitive learning neural network and based on unsupervised learning, which means that no human intervention is required during the learning process and that little needs to be known about the characteristics of the input data (Alhoniemi *et al.* 1999).

In the SOM algorithm, the topological relations and the number of neurons or nodes are fixed from the beginning. Each neuron i is represented by an n-dimensional weight, or model vector $m_i = [m_{i1},...,m_{in}]$ (n, dimension of the input vectors). Each neuron contains a weight vector. At the start of the model, the weight vectors are initialized to random values. During the training, the weight vectors are calculated using some distance measure such as the Euclidian distance, which is defined in Equation 5.1.

$$D_i = \sqrt{\sum_{j=1}^{n}(x_{ij} - m_{ij})^2} \; ; \; i = 1, 2, ..., M, \tag{5.1}$$

where

D_i = Euclidian distance between the input vector and the weight vector m;
x_{ij} = jth element of the current input vector;
m_{ij} = jth element of the weight vector m;
M = number of neurons in the SOM; and
n = dimension of the input vectors.

Node c (Equation 5.2), whose weight vector is closest to the input vector, is chosen as the best matching unit (BMU). When the BMU is found, the weight vectors m_i are updated. The BMU and its topological neighbors are moved closer to the input vector. The update rule of the weight vector is shown in Equation 5.3.

$$\|x - m_c\| = \min\{\|x - m_i\|\}, \tag{5.2}$$

where

x = input vector;
m = weight vector; and
$\| \; \|$ = a distance measure.

$$m_i(t+1) = m_i(t) + \alpha(t)h_{ci}(t)[x(t) - m_t(t)], \tag{5.3}$$

where

$m(t)$ = weight vector indicating the output unit's location in the data space at time t;
$\alpha(t)$ = learning rate at time t;
$h_{ci}(t)$ = neighborhood function centered in the winner unit c at time t; and
$x(t)$ = input vector drawn from the input data set at time t.

After this competitive learning exercise, the clusters corresponding to characteristic features can be shown on the map. The quality of the mapping is usually measured with a quantization error and a topographic error. The learning rate and neighborhood radius were set with default values. The default number of neurons was determined by the heuristic Equation 5.4. The ratio between side lengths of the map grid was set to the square root of the ratio of the two highest *eigenvalues* of the data sample (Vesanto *et al.* 2000).

$$M \approx 5\sqrt{n}, \tag{5.4}$$

where

M = number of neurons i and
n = total number of data samples.

A 2-D lattice with a map size of $M = 14 \times 7$ hexagonal units was used for both ammonia–nitrogen and SRP modeling. The final quantization and topographic errors were 8.852 and 0.096, and 6.541 and 0.123 for ammonia–nitrogen and SRP, respectively. These values were relatively low if compared to the error values with other parameter settings, indicating that the quality of the mappings was relatively good.

Since the codebook vectors of the SOM represent the local mean of the input vector, the SOM can be used for the prediction of missing components of an input vector. A prediction can be made by seeking the BMU for a vector with unknown components. The predicted values can be obtained from the BMU. The application of the SOM for prediction purposes is illustrated in Figure 5.3.

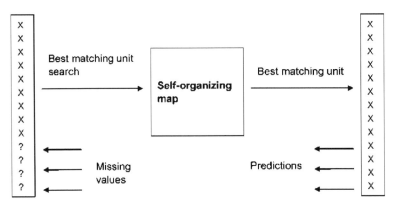

Figure 5.3 Predicting missing components of the input vector using a self-organizing map

The model is trained using the training data set, which is removed from the vector to predict a set of variables as part of an input vector. The depleted vector is subsequently presented to the SOM to identify its BMU. The values for the missing variables are then obtained by their corresponding values in the BMU (Rustum *et al.* 2008).

Lee and Scholz (2006) applied an SOM model to elucidate heavy metal removal mechanisms and to predict heavy metal concentrations in experimental constructed wetlands. The results demonstrated that heavy metals could be efficiently estimated by utilizing the SOM model.

The SOM toolbox (version 2) for Matlab 7.0 developed by the Laboratory of Computer and Information Science at Helsinki University of Technology was used in this study. The toolbox is available online at http://www.cis.hut.fi/projects/somtoolbox (Vesanto *et al.* 1999). The SOM model was applied to ammonia–nitrogen and SRP removal data to better understand the corresponding removal mechanisms in ICW.

5.3 Results and Discussion

5.3.1 Overall Performance

Mean inflow and outflow concentrations of ammonia–nitrogen and SRP are presented in Table 5.1. The reduction rates of ammonia–nitrogen and SRP were highest in 2001, because the ICW systems were young, and therefore not mature. The reduction rates were higher in the first 3 years compared to those in the following 4 years. Nevertheless, the ICW had a good treatment capacity for ammonia–nitrogen and SRP during a period of more than 6 years with removal effectiveness ranging between 97.4 and 99.2% and between 82.6 and 95.8%, respectively. In comparison, the reduction rates of ammonia–nitrogen and SRP were higher compared

Table 5.1 Mean reduction rates of ammonia–nitrogen and soluble reactive phosphorus of integrated constructed wetland systems 3, 9, and 11

Year	Ammonia–nitrogen				Soluble reactive phosphorus			
	n	In (mg/l)	Out (mg/l)	Reduction (%)	n	In (mg/l)	Out (mg/l)	Reduction (%)
2001	12	35.51	0.29	99.2	12	9.07	0.38	95.8
2002	26	38.65	0.40	99.0	27	14.67	0.67	95.4
2003	32	36.48	0.55	98.5	29	11.07	0.96	91.3
2004	54	39.85	0.55	98.6	57	13.01	1.57	87.9
2005	80	46.17	1.19	97.4	80	12.66	1.86	85.3
2006	40	32.77	0.60	98.2	32	9.47	1.65	82.6
2007	70	23.62	0.57	97.6	69	8.01	0.72	91.0

n, sample number; in, inflow; out, outflow

to those of a constructed wetland with horizontal sub-surface flow (Kyambadde *et al.* 2005): between 45.5 and 68.6% and between 45.2 and 73.5%, respectively.

The removal performances for ammonia–nitrogen were more stable in comparison to those for SRP. This corresponded well with the higher reduction rates of ammonia–nitrogen in comparison to those for SRP.

Detailed water quality results have been published by Scholz *et al.* (2007). Concerning the selected ICW systems 3, 9, and 11, the influent concentrations for ammonia–nitrogen, SRP, DO, temperature, pH, chloride, and conductivity were 36.10 mg/l, 11.14 mg/l, 5.5 mg/l, 13.7°C, 7.21, 107.9 mg/l, and 994 µS/cm, respectively, between 2001 and 2007. The corresponding outflow concentrations were 0.59 mg/l, 1.12 mg/l, 5.8 mg/l, 13.1°C, 7.51, 42.9 mg/l, and 358 µS/cm, respectively.

5.3.2 Model Application to Assess Nutrient Removal

The SOM model was applied to identify the relationships between the outflow ammonia–nitrogen concentrations and other water quality variables. The component planes for each variable of the SOM model are shown in Figure 5.4. The unified distance matrix (U-matrix) representation of the SOM visualizes the distances between the map neurons (Vesanto *et al.* 1999; Lee and Scholz 2006). The distances between the neighboring map neurons were calculated and subsequently

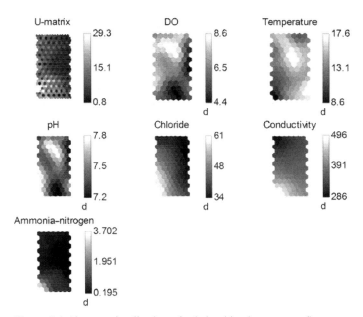

Figure 5.4 Abstract visualization of relationships between outflow ammonia–nitrogen (mg/l), and outflow dissolved oxygen (DO, mg/l), temperature (°C), pH (–), chloride (mg/l), and conductivity (µS) using a self-organizing map model

visualized by applying grey shade scaling between them; *e.g.*, the lighter grey shades are associated with the high relative component value of the corresponding weight vector. This helps to identify and subsequently illustratively show the clusters in the input data. The component plane shows the value of the variable in each map unit (Lee and Scholz 2006).

The component plane helps to visualize the relationships between ammonia–nitrogen and other variables. High outflow ammonia–nitrogen concentrations (>1.951 mg/l) are linked to high chloride concentrations (>48 mg/l), high conductivity values (>391 µS), and low temperatures (<13.1°C). Ammonia–nitrogen concentrations do not reveal an obvious association with DO concentrations and pH. Low reduction rates are apparently associated with high outflow ammonia–nitrogen concentrations as shown in Table 5.1.

High levels of conductivity and chloride represent high salt concentrations in the runoff. The linear relationship between effluent conductivity and chloride concentration is shown in Equation 5.5. Furthermore, Equations 5.6–5.8 show regression equations for ammonia–nitrogen. It can be seen that ammonia–nitrogen removal was influenced by high salt concentrations:

$$\text{Conductivity} = 3.79 \times \text{chloride} + 209.7; \ R^2 = 0.44 \text{ and } p < 0.01, \qquad (5.5)$$

$$\text{Ammonia–nitrogen} = 0.08 \times \text{chloride} - 2.6; \ R^2 = 0.15 \text{ and } p < 0.01, \qquad (5.6)$$

$$\text{Ammonia–nitrogen} = 0.02 \times \text{conductivity} - 4.9; \ R^2 = 0.18 \text{ and } p < 0.01, \quad (5.7)$$

$$\text{Ammonia–nitrogen} = 0.04 \times \text{chloride} + 0.01 \times \text{conductivity} - 4.9;$$

$$R^2 = 0.20 \text{ and } p < 0.01 \text{ (chloride and conductivity).} \qquad (5.8)$$

Chapanova *et al.* (2007) demonstrated that ammonia conversion is sensitive to the salinity of the wastewater to be treated; after adding salinity to the input wastewater, ammonia degradation was markedly reduced. However, Dincer and Kargi (1999) showed that salt concentrations >2% resulted in significant reductions in performances of both nitrification and denitrification.

In contrast, the outflow temperature is negatively correlated (R = −0.38) with the ammonia–nitrogen concentration, suggesting that temperature has a positive effect on ammonia–nitrogen removal. An elevated water temperature can enhance nitrate volatilization. Relatively high temperatures (>13.1°C) are better for both nitrification and denitrification compared to temperatures <13.1°C (USEPA 2000). Chapanova *et al.* (2007) reported that at 5°C the ammonia–nitrogen removal rate was on average three to five times smaller than at temperatures between 15 and 25°C. Nitrification is greatly affected by temperature; nitrification rates are slow in cold compared to warm climates (Chapanova *et al.* 2007; Vymazal 2007).

No obvious correlation (R = 0.10) between pH and ammonia–nitrogen could be identified. Most outflow pH values were in a range between 7.0 and 8.0 at temperatures of <17.6°C. Ammonia–nitrogen concentrations did not dminish at this pH range. Ammonia–nitrogen may be found in the un-ionized form or ionized form depending on water temperature and pH. The ionized form is predominant in wetlands; *e.g.*, at pH 7.0 and 25°C, the percentage of un-ionized ammonia is approx. 0.6% (USEPA 2000). It was also reported that at high pH ranging between 8.0 and 8.5, the proportion of ammonia might increase to between 20 and 25% at 20°C if surface turbulence is high due to wind action. Significant losses of nitrogen may occur in open water areas via ammonia gas (NH_3) volatilization (USEPA 2000; Camargo Valero and Mara 2007).

Many papers (Schaafsma *et al.* 1999; USEPA 2000; Noorvee *et al.* 2007; Iamchaturapatr *et al.* 2007) indicate that DO significantly influences the removal rate of ammonia–nitrogen in constructed wetland systems. However, DO concentrations had no obvious impact on ammonia–nitrogen removal in ICW based on the visualization of the relationship between outflow ammonia–nitrogen and DO ($R = -0.02$). Therefore, it can be seen that ammonia–nitrogen removal was largely influenced by salt concentrations and temperature. The pH and DO variables seemed to be of less importance.

A visualization of relationships between the outflow SRP concentrations and other water quality parameters of the SOM model is shown in Figure 5.5. High outflow SRP concentrations (>2.641 mg/l) are linked to high chloride concentrations (>49 mg/l) and high conductivity values (>394 µS). Unlike the case of ammonia–nitrogen removal, SRP removal was largely influenced by the DO and salt concentrations and correlated comparatively weakly with temperature and pH.

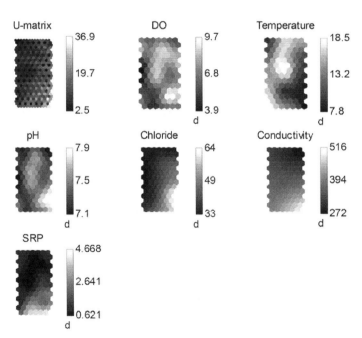

Figure 5.5 Abstract visualization of relationships between outflow soluble reactive phosphorus (SRP, mg/l) and outflow dissolved oxygen (DO, mg/l), temperature (°C), pH (–), chloride (mg/l), and conductivity (µS) using a self-organizing map model

Chloride (R = 0.48) and conductivity (R = 0.56) correlated positively with SRP, indicating that elevated salt concentrations had a negative impact on SRP removal. With increasing salt concentrations, the phosphorus removal rates of the tested ICW systems decreased. This was probably the case because phosphate accumulating microorganisms were sensitive to salinity (Scholz 2006). The salt accumulation in phosphate accumulating microorganism cells might have reached a certain threshold indicative of a significant increase in osmotic pressure in cells. Diminished phosphate accumulation capabilities subsequently result in reduced removal efficiencies as discussed previously by Carucci *et al.* (1997), and Panswad and Anan (1999).

In contrast, DO is negatively correlated (R = -0.46) with SRP, indicating that high DO concentrations had positive effects on SRP removal. The DO is an important variable influencing phosphorus removal in ICW. Case studies undertaken by Girija *et al.* (2007) revealed that the phosphorus concentrations decreased from 6.0 to 0.1 mg/l as DO concentrations increased from 0.1 to 8.6 mg/l. Low DO concentrations can cause the release of phosphorus from sediment, causing an increase of SRP (Golterman 1995; Maine *et al.* 2007).

Phosphorous might precipitate as calcium phosphate or co-precipitate with iron colloids or with calcium carbonate (Golterman 1995). For example, the USEPA (2000) reported that phosphorus might precipitate as calcium phosphate within sediment pore water or in the water column near active phytoplankton growth at pH values of >7.0. Furthermore, as pH decreases, SRP sorption to carbonates decreases while adsorption to iron increases (Golterman 1995).

Concerning the ICW study, the negative correlation (R = -0.16) between pH and SRP is weak, indicating that a high pH had a small positive influence on SRP removal in ICW. Since overall pH values were comparatively low, the influence of pH on SRP removal was weak.

The chemical composition of the three ICW systems and their effluents is complex. It follows that the key precipitation processes cannot be discussed in detail within the scope of this section. However, a detailed discussion on water quality issues with respect to an earlier, directly related, study has been published by Scholz *et al.* (2007).

Pietro *et al.* (2006) observed that phosphorus removal was weakly correlated with water temperature in a south Florida (USA) freshwater marsh. In comparison, high SRP concentrations in ICW are also associated with low temperatures (R = -0.21). However, the influence of temperature was lower for SRP removal than for ammonia–nitrogen removal.

5.3.3 *Nitrogen and Phosphorus Predictions*

The SOM model was applied to predict the ammonia–nitrogen and SRP removal performances of ICW. A correlation analysis comprising the input variables DO, temperature, pH, chloride, and conductivity and the target variables ammonia–

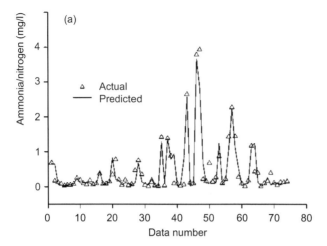

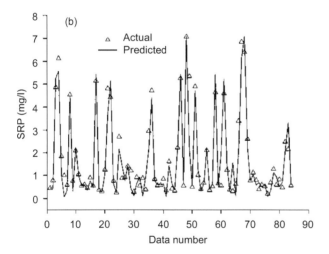

Figure 5.6 Comparing the actual and predicted outflow: (a) ammonia–nitrogen, and (b) soluble reactive phosphorus (SRP) concentrations

nitrogen and SRP was undertaken. Findings are in agreement with Figures 5.4 and 5.5 highlighting the key relationships revealed by the SOM. For example, it can be seen that ammonia–nitrogen concentrations were highly correlated with temperature, chloride, and conductivity. In comparison, SRP concentrations were highly correlated with DO, chloride, and conductivity.

In general, the measurements of input variables used for prediction should be more cost-effective, time-efficient, and easier compared with those of the target variables. Based on this consideration, temperature (R = –0.38) and conductivity (R = 0.43) were selected as input variables to predict ammonia–nitrogen in the outflow of ICW. In comparison, DO (R = –0.46) and conductivity (R = 0.56) were

Table 5.2 Summary statistics of the self-organizing map model applied for prediction purposes

Statistics	Ammonia–nitrogen prediction	Soluble reactive phosphorus prediction
Number of training data sets	240	250
Number of testing data sets	74	84
Correlation coefficient	0.934	0.951
Mean absolute scaled error	0.015	0.048

selected as input variables to predict SRP. Considering the relatively high costs and long time associated with most chloride measurement techniques, chloride was not selected for predicting either ammonia–nitrogen or SRP concentrations, even though it had comparatively strong relationships with ammonia–nitrogen (R = 0.38) and SRP (R = 0.48).

For modeling purposes, the data set was mixed in a random order and then sub-divided into two sets. The first sub-set was used as a training data set, and the second sub-set was used as a testing data set. The training and testing data sets are summarized in Table 5.2. The model was verified with the testing data set. For example, when predicting the ICW treatment performance with respect to ammonia–nitrogen removal, the corresponding ammonia–nitrogen data entries were omitted form the testing data set, implying that ammonia–nitrogen concentrations were in fact missing values.

After running the simulation, the predicted ammonia–nitrogen concentrations were subsequently compared with the actual values. The SOM modeling performances in terms of predicting the outflow ammonia–nitrogen and SRP concentrations are shown in Figure 5.6. Table 5.2 shows the summary statistics of the SOM model for the test. The SOM model has a comparatively lower mean absolute scaled error indicating its relatively high accuracy in prediction if compared to previous results (Lee and Scholz 2006). In general, the SOM model performed very well in predicting the nutrient concentrations in representative ICW systems.

5.4 Conclusions

Representative ICW were very efficient in removing ammonia–nitrogen and SRP. The SOM model showed that the ammonia–nitrogen outflow concentrations correlated with water temperature and salt concentration (indicated by conductivity and chloride). High ammonia–nitrogen removal efficiency can be achieved, if salt concentrations are low and temperatures are high. The SOM model also revealed that SRP removal was predominantly affected by salt and DO. SRP can easily be removed within ICW, if salt concentrations are low and DO, temperature, and pH values are high.

The SOM performed very well in modeling and predicting the nutrient removal in ICW. Nutrients such as ammonia–nitrogen and SRP can be accurately predicted by other more cost-effective, rapid, and easier-to-measure water quality variables such as temperature, conductivity, and DO.

References

Alhoniemi E, Hollmen J, Simula O, Vesanto J (1999) Process monitoring and modeling using the self-organizing map. Integr Comput Aided Eng 6:3–14

Allen, S.E. (1974) Chemical analysis of ecological materials. Blackwell, Oxford, UK

APHA (1998) Standard methods for the examination of water and wastewater. 20th edn. American Public Health Association (APHA), American Water Works Association and Water and Environmental Federation, Washington, DC

Astel A, Tsakovski S, Barbieri P, Simeonov V (2007) Comparison of self-organizing maps classification approach with cluster and principal components analysis for large environmental data sets. Wat Res 41:4566–4578

Camargo Valero MA, Mara DD (2007) Nitrogen removal via ammonia volatilization in maturation ponds. Wat Sci Technol 55:87–92

Carroll P, Harrington R, Keohane J, Ryder C (2005) Water treatment performance and environmental impact of integrated constructed wetlands in the Anne valley watershed, Ireland. In: Dunne EJ, Reddy KR, Carton OT (eds) Nutrient management in agricultural watersheds: a wetlands solution. Wageningen Academic Publishers, Wageningen, The Netherlands

Carty A, Scholz M, Heal K, Gouriveau F, Mustafa A (2008) The universal design, operation and maintenance guidelines for farm constructed wetlands (FCW) in temperate climates. Biores Technol 99:6780–6792

Carucci A, Majone M, Ramadori R, Rossetti S (1997) Biological phosphorus removal with different organic substrates in an anaerobic/aerobic sequencing batch reactor. Wat Sci Technol 35:161–168

Chapanova G, Jank M, Schlegel S, Koeser H (2007) Effect of temperature and salinity on the wastewater treatment performance of aerobic submerged fixed bed biofilm reactors. Wat Sci Technol 55:159–164

Dincer AR, Kargi F (1999) Salt inhibition of nitrification and denitrification in saline wastewater. Environ Technol 20:1147–1153

Girija TR, Mahanta C, Chandramouli V (2007) Water quality assessment of an untreated effluent impacted urban stream: the Bharalu Tributary of the Brahmaputra River, India. Environ Monitor Assess 130:221–236

Golterman HL (1995) The labyrinth of nutrient cycles and buffers in wetlands: results based on research in the Camargue (Southern France). Hydrobiologia 315:39–58

Hammer DA, Bastian RK (1989) Wetlands ecosystems: natural water purifiers. In: Hammer DA (ed) Constructed wetlands for wastewater treatment. Lewis, Chelsea, MI

Harrington R, Ryder C (2002) The use of integrated constructed wetlands in the management of farmyard runoff and waste water. National Hydrology Seminar on Water Resources Management Sustainable Supply and Demand. The Irish National Committees of the International Hydrological Programme and the International Commission on Irrigation and Drainage, Tullamore, Ireland

Harrington R, Dunne EJ, Carroll P, Keohane J, Ryder C (2005) The concept, design and performance of integrated constructed wetlands for the treatment of farmyard dirty water. In: Dunne EJ, Reddy KR, Carton OT (eds) Nutrient management in agricultural watersheds: a wetlands solution. Wageningen Academic Publishers, Wageningen, The Netherlands

Iamchaturapatr J, Yi SW, Rhee JS (2007) Nutrient removals by 21 aquatic plants for vertical free surface-flow (VFS) constructed wetland. Ecol Eng 29:287–293

Kohonen T, Oja E, Simula O, Visa A, Kangas J (1996) Engineering applications of the self organizing map. Proc IEEE 84:1358–1384

Kyambadde J, Kansiime F, Dalhammar G (2005) Nitrogen and phosphorus removal in substrate-free pilot constructed wetlands with horizontal surface flow in Uganda. Wat Air Soil Pollut 165:37–59

Lee B-H, Scholz M (2006) Application of the self-organizing map (SOM) to assess the heavy metal removal performance in experimental constructed wetlands. Wat Res 40:3367–3374

Maine MA, Suñé N, Hadad HR, Sánchez G (2007) Temporal and spatial variation of phosphate distribution in the sediment of a free surface water constructed wetland. Sci Total Environ 380:75–83

Noorvee A, Poldvere E, Mander Ü (2007) The effect of pre-aeration on the purification processes in the long-term performance of a horizontal subsurface flow constructed wetland. Sci Total Environ 380:229–236

Panswad T, Anan C (1999) Impact of high chloride wastewater on an anaerobic/anoxic/aerobic process with and without inoculation of chloride acclimated seeds. Wat Res 33:1165–1172

Pietro KC, Chimney MJ, Steinman AD (2006) Phosphorus removal by the *Ceratophyllum*/periphyton complex in a south Florida (USA) freshwater marsh. Ecol Eng 27:290–300

Rustum R, Adeloye JA, Scholz M (2008) Applying Kohonen self-organizing map as a software sensor to predict biochemical oxygen demand. Wat Environ Res 80:32–40

Schaafsma JA, Baldwin AH, Streb CA (1999) An evaluation of a constructed wetland to treat wastewater from a dairy farm in Maryland, USA. Ecol Eng 14:199–206

Scholz M (2006) Wetland systems to control urban runoff. Elsevier, Amsterdam

Scholz M (2008) Classification of flood retention basins: The Kaiserstuhl case study. Environ Eng Geosci 24:61–80

Scholz M, Harrington R, Carroll P, Mustafa A (2007) The integrated constructed wetlands (ICW) concept. Wetlands 27:337–354

US EPA (2000) Constructed wetlands treatment of municipal wastewater. United States (US) Environmental Protection Agency (EPA), Office of Research and Development, Cincinnati, OH, USA

Vesanto J, Alhoniemi E (2000) Clustering of the self-organizing map. IEEE Trans Neural Netw 11:586–600

Vesanto J, Himberg J, Alhoniemi E, Parhankangas J (1999) Self-organizing map in Matlab: the SOM toolbox. In: Proceedings of the Matlab DSP Conference, November 1999, Espoo, Finland, pp. 34–40. Software available online at http://www.cis.hut.fi/projects/somtoolbox/. Accessed 10 Jan 2010

Vesanto J, Himberg J, Alhoniemi E, Parhankangas J (2000) SOM toolbox for Matlab 5 documentation. Helsinki University of Technology, Helsinki, Finland. Software available online at http://www.cis.hut.fi/projects/somtoolbox/. Accessed 10 Jan 2010

Vymazal J (2007) Removal of nutrients in various types of constructed wetlands. Sci Total Environ 380:48–65

Index